Thermal Energy Battery with Nano-enhanced PCM

Edited by
Mohsen Sheikholeslami Kandelousi

Published in London, United Kingdom

IntechOpen

Supporting open minds since 2005

Thermal Energy Battery with Nano-enhanced PCM
http://dx.doi.org/10.5772/intechopen.75858
Edited by Mohsen Sheikholeslami Kandelousi

Contributors
Tatang Mulyana, Getu Hailu, Zhiguo Qu, Jack Gorelik, Artur Khabitov, Mohsen Sheikholeslami Kandelousi

First published in London, United Kingdom, 2019 by IntechOpen
IntechOpen is the global imprint of INTECHOPEN LIMITED, registered in England and Wales, registration number: 11086078, The Shard, 25th floor, 32 London Bridge Street
London, SE19SG – United Kingdom
Printed in Croatia

British Library Cataloguing-in-Publication Data
A catalogue record for this book is available from the British Library

Additional hard and PDF copies can be obtained from orders@intechopen.com

Thermal Energy Battery with Nano-enhanced PCM
Edited by Mohsen Sheikholeslami Kandelousi
p. cm.
Print ISBN 978-1-78985-417-6
Online ISBN 978-1-78985-418-3
eBook (PDF) ISBN 978-1-83962-178-9

For EU product safety concerns:
IN TECH d.o.o., Prolaz Marije Krucifikse Kozulić 3, 51000 Rijeka, Croatia, info@intechopen.com or visit our website at intechopen.com.

Meet the editor

Dr. Mohsen Sheikholeslami Kandelousi works at the Babol Noshirvani University of Technology's Department of Mechanical Engineering in Iran. He is head of the Renewable Energy Systems and Nanofluid Applications in Heat Transfer Laboratory at Babol Noshirvani University of Technology. His research interests are nanofluids, computational fluid dynamics, simulation, mesoscopic modeling, nonlinear science, magnetohydrodynamics, ferrohydrodynamics, electrohydrodynamics, and heat exchangers. He has written several papers and books in various fields of mechanical engineering. He is the first scientist to develop a new numerical method (control-volume finite-element method) and he has published a reference book *Application of Control Volume Based Finite Element Method (CVFEM) for Nanofluid Flow and Heat Transfer*. He is also the first author of the books *Applications of Nanofluid for Heat Transfer Enhancement*, *Application of Semi Analytical Methods for Nanofluid Flow and Heat Transfer*, *Hydrothermal Analysis in Engineering Using Control Volume Finite Element Method*, and *External Magnetic Field Effects on Hydrothermal Treatment of Nanofluid*, published by Elsevier. According to the reports of Thomson Reuters (Clarivate Analytics), he was selected as a Web of Science Highly Cited Researcher (Top 0.01%) in 2016, 2017, and 2018.

Contents

Preface

Nanoparticle-enhanced phase change materials have engrossed augmented consideration to remove the main limitation of phase change materials (PCMs) in various industrial uses. In this book, a number of features of nanoparticle-enhanced phase change materials (NEPCMs) are provided. Using nanoparticles intensifies energy performance and effective thermal management. Overviews of recent work and developments in the use of NEPCMs are presented in Chapter 1 (Introductory chapter). Nanoparticles are employed to expedite unsteady processes and improve performance. Usually, the amount of nanoparticles is less than 5% by weight. In this case, the impact of adding nanoparticles on PCMs is investigated. In Chapter 2, the influence of solar energy on energy storage systems is studied. Solar energy is one of the most important renewable energy sources available in most parts of the world. One of the limitations of using solar energy is that it cannot be used at night, and the amount of solar light that the globe receives depends on factors such as location, time of day, time of year, and weather conditions. The lattice Boltzmann method is presented in Chapter 3 to simulate heat transfer enhancement of PCMs. Experimental studies on the cooling of industrial machines are reported in Chapter 4. Chapter 5 deals with a new model for designing heat exchangers.

Dr. Mohsen Sheikholeslami Kandelousi
Department of Mechanical Engineering,
Babol Noshirvani University of Technology,
Babol, Iran

Renewable Energy Systems and Nanofluid Applications in
Heat Transfer Laboratory, Babol Noshirvani University of Technology,
Babol, Iran

Chapter 1

Introductory Chapter: Nano-Enhanced Phase-Change Material

Mohsen Sheikholeslami Kandelousi

1. Phase-change materials

The consumption of any kind of energy has a significant role in protecting energy in the economic development of any country. Today, request in the sector has led to beautiful and large buildings around the world. It is noteworthy that buildings will spend about 30% of the worldwide energy produced. More energy demand for developing economies is projected; the lack of awareness about the pricing of fossil fuels and the increasing information about environmental problems has donated to the serious consideration of different renewable energies. Among different types of energy, heat transfer is universally observed in nature as geothermal energy and energy radiation in nature. This energy is a side effect of a large amount of energy production units and energy conversion products and systems. Energy storage may be maintained to maintain energy before energy use, or it can be used as a way to maintain calm in buildings, saving energy in different sectors of the society. An energy storage system should have certain features that include proper energy storage material with a specific melting temperature at the optimum range, decent heat transfer well, and pleasant enclosure compatible with the most important energy storage methods. Today, to reduce fossil fuel consumption in buildings, thermal energy storage systems should be employed to augment the efficiently. **Figure 1** shows examples for the industrial and internal application of thermal energy storage systems (see [1, 2]). One of the advantages of utilizing thermal storage is that it can be consistent with supply [3].

2. Techniques for energy storage

2.1 Latent heat

Release or absorption of energy by changing the phase as gas to liquid or liquid to solid can be considered as the main reason for such way.

2.2 Sensible heat

Energy is saved by increasing material temperature. Efficiency of thermal storage relies on temperature difference, specific heat, and amount of material. The best material in this view is water due to its low cost and great heat capacity.

IntechOpen

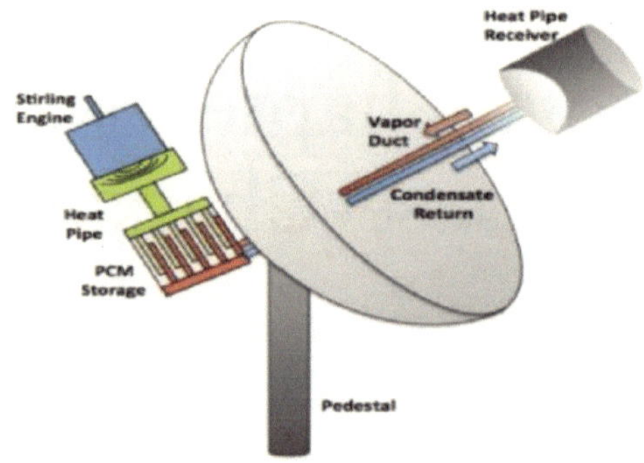

a)

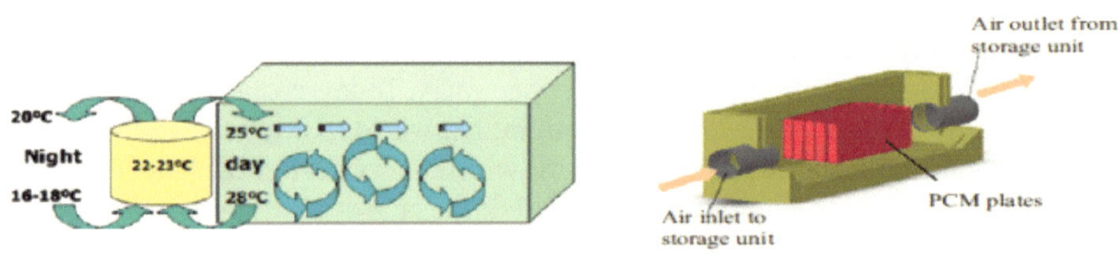

b)

Figure 1.
NEPCM for energy storage units [1, 2]. a) Latent energy storage for dish striling power generation b) PCM in free cooling systems.

2.3 Electrical energy

The electric energy has been stored through the battery. Battery is charged directly to a direct current source and converted to the electrical energy when discharged. Storing of electrical energy is produced by photovoltaic cells with the use of battery during the power loss.

2.4 Mechanical energy

This system includes the storage of gravitational energy, pumped water power storage, and the storage of compressed air energy.

2.5 Thermochemical energy storage

Absorption and discharge of energy through the failure and deformation of the molecular bond during the chemical reaction can be considered as the basis of this method. The mass of material, the type of reaction, and the amount of change affect the stored heat directly. The phase variation can take the form of solid-gas, liquid-gas, solid-solid, and solid-liquid. In change of solid, the heat is saved in a change from one type of crystal to another. These changes generally have low-energy content and a small change in volume relative to the change of solid to liquid. Heat transfer takes place when the phase changes. Liquid phase and solid phase converge up and down by absorbing and releasing heat at temperatures that release or absorb heat at the uniform temperatures. These materials store energy (5–14 times) relative to tangible energy storage. Phase-to-phase materials are more than loneliness for the transfer of heat, but to use this material, there is a need for a heat exchanger between the PCM and source. This is because of the low diffusion factor of PCM.

2.6 Heat storage

Heat storage is a variation of internal energy of the material which can be sensible, latent, and thermochemical forms.

3. PCM

3.1 Categories

Phase materials are organic or mineral compounds can save and absorb high heat energy within them. The storage of heat energy in these materials occurs within phase-change mechanism. In distinctive sources, different categories for PCM were performed, but as the common area of these divisions, mainly phase-change materials fall into two nonorganic categories. The most common organic material of this material is paraffin wax and the most common nonorganic sample of hydrate salt. It is worth noting that there are significant differences between the characteristics of organic phase transition material and nonorganic PCMs. Three important categories must be categorized into hydrate salts, organic matter, and eutectics. Organic materials are also composed of high-carbon and hydrogen chains whose phase-change temperatures are above 0°C (see **Figure 2**). Comparison of the cases is listed in **Table 1**. In the following section, varied chemistry categories of phase-change materials, disadvantages, and advantages are discussed. The criteria for choosing a suitable phase-change material for different uses are discussed below. Then, to define PCMs, plus some examples of applications of these materials, techniques for increasing the efficiency of these materials in energy storage systems have been investigated. Finally, with a review of past work in this area, a summary of the whole chapter is presented.

3.1.1 Inorganic materials

The inorganic PCMs are generally categorized into two groups of metals and hydrated salts. These phase-change materials are denser and more energy-efficient than organic phase-change materials. In terms of price, organic phase-change materials are extensively cheaper. In addition to the benefits mentioned above, inorganic phase-change materials are widely corrosive. Generally, the problem is overcooling, which means they are hardly crystalline when they reach the freezing point.

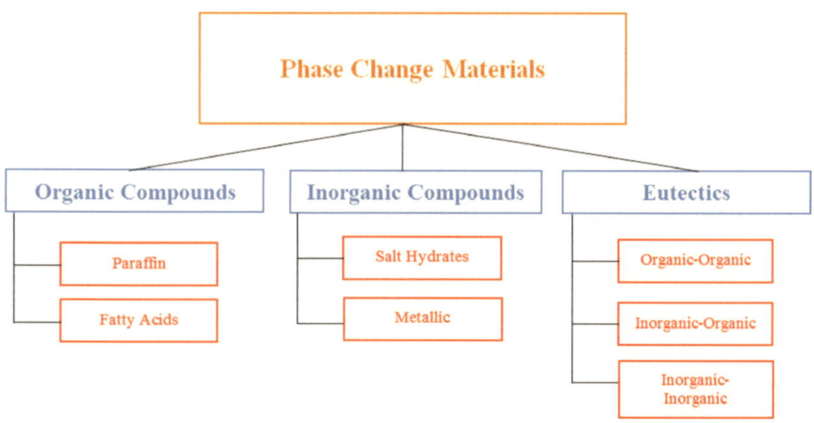

Figure 2.
Different kinds of PCM.

Disadvantages	Benefits	Classification
Corrosion Supercoiling	Availability in low cost Low volume change High thermal conductivity High heat of fusion	Inorganic PCMs
Flammability Low thermal conductivity Relative large volume change	Good compatibility Recyclable with other materials No supercoiling High heat of fusion Availability in a large temperature range	Organic PCMs
Lack of test data	High volumetric thermal storage density Sharp melting temperature	Eutectics

Table 1.
Various types of PCMs.

3.1.2 Organic materials

The organic PCMs should be separated from the group of paraffin and non-paraffins, such as fatty acids, esters, glycolic acids, and alcohols.

3.1.3 Disadvantages and advantage of organic and nonorganic material

The advantages of organic material include the presence in the temperature range, freezing of overcooling, ability to melt compatible with common materials, non-separable and recyclable, and great fusion heat. The limitations of this organic material include low storage capacity, low thermal conductivity, and thermal flammability. The advantages of nonorganic materials include considerable latent heat and low cost. The limitations of nonorganic material include a large change in volume, overheating, separation, and corrosion.

3.1.4 Phase-change material selection and selection criteria

To select the phase-change material, there exist some parameters that should be studied, namely, thermodynamic characteristics, kinematic properties, chemical properties, and economic features.

3.1.4.1 Thermodynamic characteristics

The melting temperature must react in the wanted temperature range and low vapor pressure at the reaction temperature to decrease the issue of inhibiting melting.

3.1.4.2 Kinematic properties

The high reaction rate is important for avoiding the superconducting fluid phase and high growth rates; therefore, system can restore heat from storage units.

3.1.4.3 Economic features

It is economically low-cost and also available.

3.1.4.4 Chemical properties

The chemical and full cycle and non-degradation after the successive cycles are important issues.

3.1.5 Application of phase-change material

The materials of phase changes have found many applications due to the changing temperature. Materials that melt below 15°C can be used to cool and ventilate room air are used. Materials that melt above 90°C are used to reduce temperatures where temperatures can rise suddenly and prevent fire. The PCM, whose melting temperature is between these two values, is used to store solar energy. PCMs for cold and heating temperatures in smaller proportions have building applications. In addition to the melting point of PCM, in the design of each heat storage unit that operates on the basis of PCM, the following notes must be considered: the material of the desired PCM with the wanted melting temperature and a PCM storage chamber that is talented of absorbing the volume changes of these materials within phase change. It should be noted that in order to get better efficiency, the melting temperature should be within the working temperature range, in addition to its thermal insulation, the phase-change properties can also be used. However, there are plenty of ways to use these materials in design of buildings. But one of the most significant points in the use of such materials, like other static systems, is the issue of calculating economic justification and the time of returning capital after the time of operation. In some cases, the use of PCM can decrease the thermal flux of the entrances to the building by up to 38%. Furthermore, the augment in the number of holes in the PCM in the recipe makes a reduction of 11% in heat flux. However, since the use of these materials in raw form can be disadvantageous due to the type PCM, and also due to the possibility of fluid PCMs, it can be necessary to enclose in the liquid phase. In order to provide both micro- and macrocomponents, phase changes are embedded in the enclosures and are thus encapsulated. In this way, the major difference between microcapsules and microcapsules is in the size of the casings or capsules. In the microcapsule method, the materials are embedded in spheres of 1–30 microns in diameter. In microcapsule method, materials in larger compartments are embedded in envelopes or containers of different sizes. The bulk of these compartments are made of plastic packs, as well as high-density polyethylene panels. Both microcapsules and microscopes have advantages and disadvantages. Encapsulation of materials is carried out using various materials. Encapsulation using plastics and metals is expensive but safe. In the construction industry, both methods can be used, but according to the place of use and also in combination with other building materials, the application of each method creates significant differences among the results and thermal treatment of the building. In the application of microcapsules of phase-change materials as fine-grained concrete, attention to the lack of functional interactions between these materials is important for the initial performance of fine-grained concrete. In this way, micro-capsulation of the material by saturation of the phase change in the fresh concrete is very effective in the thermal behavior of the materials.

3.1.5.1 PCM for heat storage

Newly, several researchers focus on improving the efficiency of building energy in order to enhance the performance of buildings [4]. Among different technologies, PCM has a major contribution to the development of high-efficiency buildings [5].

PCMs are materials that are talented of high absorption, release, and storage of energy at a relatively stable temperature. Thermal energy can be saved by saving energy or sensible energy with thermal changes in material content. This energy is available when the process is reversed. A number of papers have been reviewed [6] on PCMs that have been investigated for heat transfer, transportation, and various issues related to the system. Due to high-density changes, phase-change materials that require supports are not preferred over the high-energy process. Thus, for thermal applications, the latent fusion of PCMs is used. For a specific program, melting point and latent heat are important criteria to be chosen. Many of the phases modifying substances are considered, such as paraffin, water, and salt hydrates. Some properties, including diffusivity and thermal conductivity, are very important [7]. To improve the storage of thermal energy, the first review of nanomaterials in phase-modified materials is given by Elgafy and Lafdi [8]. Some effective applications of phase-change materials have been presented by Salyer et al. [9] and Demirbas [10], which can be referred to solar power plants for saving daytime energy and reuse it the next day, medical treatment, pharmaceuticals, pharmaceuticals, transportation, photovoltaic cells, thermal management systems, and electronic systems to prevent very high temperatures. One of the main uses of this material in the building is energy conservation.

4. Nanotechnology

4.1 Use of nanotechnology in heat storage unit

Development of nanotechnology has led to the emergence of a bunch of high-performance nano-structured materials. These materials are capable of absorbing thermal energy and high release and become for industrial applications. NEPCM has been used for various applications [11–36].

Author details

Mohsen Sheikholeslami Kandelousi[1, 2]

1 Department of Mechanical Engineering, Babol Noshirvani University of Technology, Babol, Iran

2 Renewable Energy Systems and Nanofluid Applications in Heat Transfer Laboratory, Babol Noshirvani University of Technology, Babol, Iran

*Address all correspondence to: mohsen.sheikholeslami@yahoo.com

References

[1] Andraka CE, Kruizenga AM, Hernandez-Sanchez BA, Coker EN. Review on phase change materials with nanoparticle in engineering applications. Journal of Engineering Science and Technology Review. 2016;**9**:26-386

[2] Zalba B, Marín JM, Cabeza LF, Mehling H. Free-cooling of buildings with phase change materials. International Journal of Refrigeration. 2004;**27**:839-849

[3] Sharma SD, Sagara K. Latent heat storage materials and systems, a review. International Journal of Green Energy. 2005;**2**:1-56

[4] Hughes BR, Chaudhry HN, Ghani SA. A review of sustainable cooling technologies in buildings. Renewable and Sustainable Energy Reviews. 2011;**15**:3112-3120

[5] Zhu N, Ma Z, Wang S. A review of dynamic characteristics and energy performance of buildings using phase change materials. Energy Conversion and Management. 2009;**50**:3169-3181

[6] Abhat A. Low temperature latent heat thermal energy storage-heat storage materials. Solar Energy. 1983;**30**:313-332

[7] Fleischer SA. Thermal Energy Storage Using Phase Change Materials, Fundamentals and Applications. USA: Springer; 2015

[8] Elgafy A, Lafdi K. Effect of carbon nanofiber additives on thermal behavior of phase change materials. Carbon. 2005;**43**:3067-3074

[9] Salyer IO, Sircar K. A review of phase change materials research for thermal energy storage in heating and cooling applications at the University of Dayton from 1982 to 1996. International Journal of Global Energy Issues. 1997;**9**:183-198

[10] Demirbas F. Thermal energy storage and phase change materials, an overview. Energy Sources. 2006;**1**:85-95

[11] Sheikholeslami M. Finite element method for PCM solidification in existence of CuO nanoparticles. Journal of Molecular Liquids. 2018;**265**:347-355

[12] Sheikholeslami M, Ghasemi A, Li Z, Shafee A, Saleem S. Influence of CuO nanoparticles on heat transfer behavior of PCM in solidification process considering radiative source term. International Journal of Heat and Mass Transfer. 2018;**126**:1252-1264

[13] Sheikholeslami M. Solidification of NEPCM under the effect of magnetic field in a porous thermal energy storage enclosure using CuO nanoparticles. Journal of Molecular Liquids. 2018;**263**:303-315

[14] Sheikholeslami M. Numerical simulation for solidification in a LHTESS by means of nano-enhanced PCM. Journal of the Taiwan Institute of Chemical Engineers. 2018;**86**:25-41

[15] Sheikholeslami M, Ghasemi A. Solidification heat transfer of nanofluid in existence of thermal radiation by means of FEM. International Journal of Heat and Mass Transfer. 2018;**123**:418-431

[16] Sheikholeslami M, Zeeshan A, Majeed A. Control volume based finite element simulation of magnetic nanofluid flow and heat transport in non-Darcy medium. Journal of Molecular Liquids. 2018;**268**:354-364

[17] Zhixiong Li M, Sheikholeslami MS, Shafee A. Nanofluid unsteady heat transfer in a porous energy storage enclosure in existence of Lorentz forces. International Journal of Heat and Mass Transfer. 2018;**127**:914-926

[18] Sheikholeslami M. Application of Darcy law for nanofluid flow in a porous cavity under the impact of Lorentz forces. Journal of Molecular Liquids. 2018;**266**:495-503

[19] Li Z, Sheikholeslami M, Shafee A, Saleem S, Chamkha AJ. Effect of dispersing nanoparticles on solidification process in existence of Lorenz forces in a permeable media. Journal of Molecular Liquids. 2018;**266**:181-193

[20] Sheikholeslami M, Li Z, Shafee A. Lorentz forces effect on NEPCM heat transfer during solidification in a porous energy storage system. International Journal of Heat and Mass Transfer. 2018;**127**:665-674

[21] Sheikholeslami M. Numerical modeling of nanoenhanced PCM solidification in an enclosure with metallic fin. Journal of Molecular Liquids. 2018;**259**:424-438

[22] Sheikholeslami M, Sadoughi MK. Simulation of CuO-water nanofluid heat transfer enhancement in presence of melting surface. International Journal of Heat and Mass Transfer. 2018;**116**:909-919

[23] Sheikholeslami M, Rokni HB. Melting heat transfer influence on nanofluid flow inside a cavity in existence of magnetic field. International Journal of Heat and Mass Transfer. 2017;**114**:517-526

[24] Sheikholeslami M. Numerical investigation for CuO-H_2O nanofluid flow in a porous channel with magnetic field using mesoscopic method. Journal of Molecular Liquids. 2018;**249**:739-746

[25] Sheikholeslami M, Rokni HB. CVFEM for effect of Lorentz forces on nanofluid flow in a porous complex shaped enclosure by means of non-equilibrium model. Journal of Molecular Liquids. 2018;**254**:446-462

[26] Sheikholeslami M. Numerical investigation of nanofluid free convection under the influence of electric field in a porous enclosure. Journal of Molecular Liquids. 2018;**249**:1212-1221

[27] Sheikholeslami M. Influence of magnetic field on Al_2O_3-H_2O nanofluid forced convection heat transfer in a porous lid driven cavity with hot sphere obstacle by means of LBM. Journal of Molecular Liquids. 2018;**263**:472-488

[28] Sheikholeslami M. CuO-water nanofluid flow due to magnetic field inside a porous media considering Brownian motion. Journal of Molecular Liquids. 2018;**249**:921-929

[29] Sheikholeslami M, Darzi M, Li Z. Experimental investigation for entropy generation and exergy loss of nano-refrigerant condensation process. International Journal of Heat and Mass Transfer. 2018;**125**:1087-1095

[30] Zhixiong Li, Sheikholeslami M, Shafee A, Rizwan-ul H, Khan I, Tlili I, Kandasamy R. Solidification process through a solar energy storage enclosure using various sizes of Al_2O_3 nanoparticles. Journal of Molecular Liquids. 2019. https://doi.org/10.1016/j.molliq.2018.11.129

[31] Ramzan M, Sheikholeslami M, Chung JD, Shafee A. Melting heat transfer and entropy optimization owing to carbon nanotubes suspended Casson nanoliquid flow past a swirling cylinder-A numerical treatment. AIP ADVANCES. 2018;**8**:115130. https://doi.org/10.1063/1.5064389

[32] Sheikholeslami M, Rizwan-ul H, Shafee A, Zhixiong L. Heat transfer behavior of Nanoparticle enhanced PCM solidification through an enclosure with V shaped fins. International Journal of Heat and Mass Transfer. 2019;**130**:1322-1342

[33] Sheikholeslami M. New computational approach for exergy and

entropy analysis of nanofluid under the impact of Lorentz force through a porous media. Computer Methods in Applied Mechanics and Engineering. 2019;**344**:319-333

[34] Sheikholeslami M. Numerical approach for MHD Al_2O_3-water nanofluid transportation inside a permeable medium using innovative computer method. Computer Methods in Applied Mechanics and Engineering. 2019;**344**:306-318

[35] Sheikholeslami M, Barzegar Gerdroodbary M, Moradi R, Shafee A, Zhixiong L. Application of Neural Network for estimation of heat transfer treatment of Al_2O_3- H_2O nanofluid through a channel. Computer Methods in Applied Mechanics and Engineering. 2019;**344**:1-12

[36] Zhixiong L, Sheikholeslami M, Shafee A, Ramzan M, Kandasamy R, Al-Mdallal QM. Influence of adding nanoparticles on solidification in a heat storage system considering radiation effect. Journal of Molecular Liquids. 2019;**273**:589-605

Chapter 2

Seasonal Solar Thermal Energy Storage

Getu Hailu

Abstract

Solar intermittency is a major problem, and there is a need and great interest in developing a means of storing solar energy for later use when solar radiation is not available. Thermal energy storage (TES) is a technology that is used to balance the mismatch in demand and supply for heating and/or cooling. Solar thermal energy storage is used in many applications: buildings, concentrating solar power plants and industrial processes. Solar thermal water heaters capable of heating water during the day and storing the heated water for evening use are common. TES improves system performance by smoothing supply and demand and temperature fluctuations. Thermal energy storage has become a fast-growing business. According to a research report, the global thermal energy storage market is expected to reach USD 12.50 billion by 2025. The chapter describes different types of thermal energy storage systems. Brief history, current state of research and the future of thermal storage are presented. Types of thermal storages, classifications, advantages and disadvantages are discussed; important thermal and physical properties are tabulated. Advances in enhancement of thermal properties of materials are briefly discussed. Challenges, opportunities, market outlook, government incentives and polices that support deployment of energy storage systems are outlined.

Keywords: thermal energy storage, sensible heat storage, latent heat storage, phase change materials (PCM), energy storage

1. Introduction

Solar thermal energy storage is not a new concept. Early humans had realized the abundance of solar energy and devised many methods of capturing that energy. The Greek historian Xenophon wrote of the teachings of Socrates on how to orient a building as to keep it warm in the winter and cool in the summer. Romans would place many windows on a bath houses' south wall to provide heating for their baths and reduce the fuel needed for their hypocaust, or bath fire. Native Americans in the canyons of Arizona used the southern cliff exposure of a canyon to heat their adobe buildings cleverly placed in caves just so that the low winter Sun angle would soak them with sunlight while the summer angle would be higher and therefore missing the buildings [1].

Thermal energy storage dates to the times when humans lived in natural caves. Caves are warm in winter and cold in summer when compared to the outside temperature. Cave dwellers took advantage of deep underground caves (deep underground structures), which have almost negligible temperature variations with season. Historical records show that the oldest form of thermal energy storage is harvesting ice and snow for food preservation, cold drinks and space cooling [2]. Historical records also show that Romans, Greeks and Chinese explored the use of curved mirrors to concentrate the Sun's rays that could cause flames and explosion. According to a Greek legend, in 212 B.C., Archimedes used mirrors to focus sunlight on ships of an invading Roman fleet at Syracuse and destroyed the fleet [3].

In 1767, a French-Swiss scientist Horace Benedict de Saussure built the first solar heat trapper that could be used for cooking [4]. More than 100 years later, in 1876, Adams invented the first mass-produced solar thermal device, by adding solar energy concentrator to de Saussure's solar heat trapper [5, 6]. Adams' octagonal solar oven equipped with eight solar concentrators (mirrors), reportedly cooked rations for seven soldiers in 2 h. Adams reported, "The rations of seven soldiers, consisting of meat and vegetables, are thoroughly cooked by it in two hours, in January, the coldest month of the year in Bombay, and the men declare the food to be cooked much better than in the ordinary manner." His solar oven was mass produced in India and became quite popular. In the United States, the Adams' solar oven had become a popular product for camping and an educational device for teenage students [3, 6]. Cooking for soldiers using solar oven was also a concept a French mathematician Mouchot investigated. A year after Adams' invention, in 1877, Mouchot devised solar cookers for French soldiers in Algeria, including a shiny metal cone, made from a 105.5° section of a circle. He built a separate cooker to steam vegetables and wrote the first book on Solar Energy and its industrial applications [7]. Wang Xiping, a participant of the First World Conference in Solar Cooking, held in Stockton, California, in 1992, presented the use of solar cooking in China, with Peking duck roasted in Xiao's Duck Shop, Chengdu, China, in 1894 [8].

Clarence Kemp was credited with the first commercially available solar water heater patented in 1891 called "Climax" [9]. The invention of the Climax led to a brief explosion in solar thermal technologies. William Baily in 1909 developed a solar water heater called the "Day and Night," where the heater is separated from an insulated tank to extend the duration of available hot water [9]. The first commercial solar power unit produced steam and was built in Egypt in 1913, with its energy cost being compared to coal at the time [10]. However, these earlier technologies relied on the daily solar cycles and could only be used in climates where ambient temperatures remained above freezing.

Thermal energy storage has now become a fast-growing business. "I expect thermal will be bigger than batteries," said Ice Energy CEO Mike Hopkins, "because thermal loads are the large loads. They are the problematic loads; they are the loads that don't lend themselves to using electrical storage." [11]. Today, most of the thermal energy storage technologies have either been fully developed and commercialized or are in the demonstration and development stage. **Figure 1** shows stages of different thermal storage technologies. In **Figure 1**, some key technologies are displayed with respect to their associated initial capital investment requirements and technology risk versus their current phase of development (i.e., R & D, demonstration and deployment or commercialization phases) [12].

Current research and development of thermal energy storage is mainly focused on reducing the costs of high-density storage, including thermochemical process and phase-change material (PCM) development [12]. Thermal storage systems have found applications worldwide. For example, cold-water storage tanks have been installed around the world to supply cooling capacity in commercial and industrial settings. In Canada, Germany, the Netherlands and Sweden, borehole and aquifer thermal energy storages provide both heating and cooling. In the United States, an estimated 1 GW of ice storage has been deployed to reduce peak energy consumption in areas with high numbers of cooling-degree days [13]. Borehole and aquifer

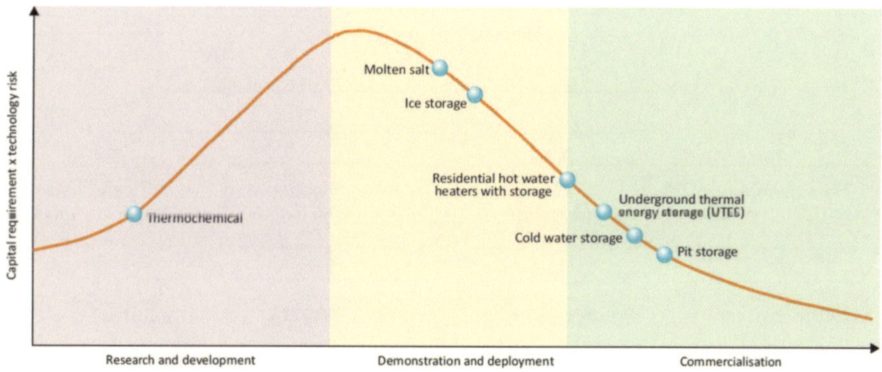

Figure 1.
Maturity of thermal energy storage technologies [12].

systems have been successfully deployed on a commercial scale to provide heating capacity in the Netherlands, Norway and Canada. The Drake Landing Solar Community in Okotoks, Canada is the first major implementation of borehole seasonal thermal energy storage in district heating in North America. It is also the first system of this type designed to supply more than 90% space heating with solar energy and the first operating in such a cold climate [14].

Thermochemical storage, in which reversible chemical reactions are used to store cooling capacity in the form of chemical compounds, is currently a focus in thermal storage R&D projects due to its ability to achieve energy storage densities of 5–20 times greater than sensible storage [12].

The focus of the following section will be on sensible and latent heat thermal storage technologies (i.e., technologies that have been fully developed and commercialized or are in the demonstration and deployment stage).

2. Types of thermal energy storage technology

Thermal energy storage is a technology that allows the transfer of heat and storage in a suitable medium. It is a technology that allows storage of thermal energy by heating or cooling a storage medium for a later use for either heating or cooling applications and power generation. Seasonal storage is defined as the ability to store energy for days, weeks or months to compensate for a longer term supply disruption or seasonal variability on the supply and demand sides of the energy system (e.g., storing heat in the summer for use in the winter via underground thermal energy storage systems) [12]. Advantages of using thermal energy storages include: increased overall efficiency, better reliability, better economics and less pollution of the environment (lower carbon dioxide - CO_2) emissions [15]. The selection of thermal energy storage depends on the required storage duration, that is, diurnal or seasonal, economic viability, the type of energy source and operating conditions. Thermal energy storages can be classified based on the type of application, type of end user, type of technology and by the type of storage material used. **Figure 2** shows the classification in detail.

The major characteristics of a thermal energy storage system are (a) its capacity per unit volume; (b) the temperature range over which it operates, that is, the temperature at which heat is added to and removed from the system; (c) the means of addition or removal of heat and the temperature differences associated therewith; (d) temperature stratification in the storage unit; (e) the power requirements

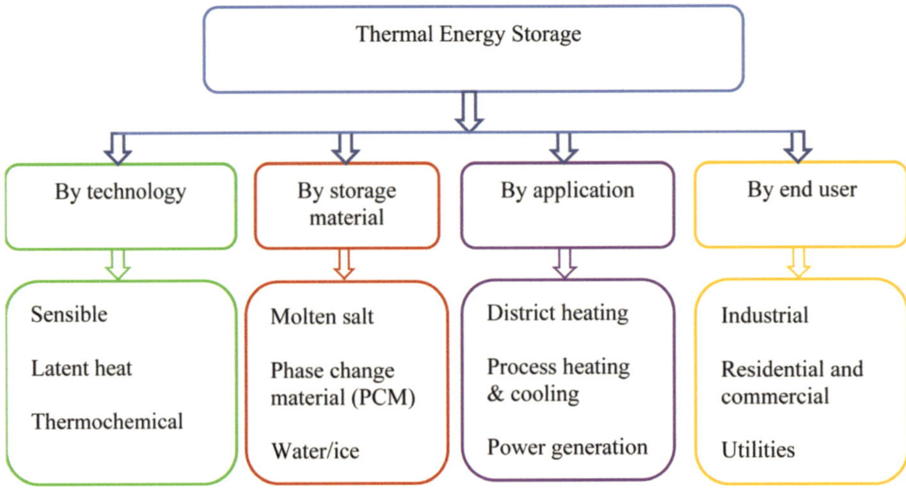

Figure 2.
Classification of thermal energy storage systems.

for addition or removal of heat; (f) the containers, tanks or other structural elements associated with the storage system; (g) the means of controlling thermal losses from the storage system and (h) its cost [16].

2.1 Sensible heat storage system

The most common type of thermal energy storage is sensible heat storage which utilizes both solid and liquid types of storage medium such as rock, sand, clay, earth, water and oil. In sensible heat storage change in temperature of the medium occurs, that is, the temperature is either increased or decreased. Heat is removed from the storage whenever required to satisfy a load, such as space heating or for domestic hot water. The removal of heat from the storage lowers its temperature. Although there are many possibilities of variations, sensible heat storage medium always consists of: an insulated container, heat storage medium and means for adding and removing heat.

In sensible hot heat storage systems, heat is added (i.e., the temperature is increased) to the storing medium, whereas in sensible cold storage systems, heat is removed thus lowering the temperature. In SHS system, the amount of energy stored is proportional to the difference between the storage medium's input and output temperatures, the mass of the storage medium and the medium's heat capacity [17]. The fundamental equation for calculating the amount of heat stored for sensible heat storage systems (SHS) is

$$Q = mC_p\Delta T = \rho V C_p \Delta T \tag{1}$$

where Q is the amount of heat stored [J], m is mass of the storage material [kg], C_p is specific heat of the storage material [J/kg K], ΔT is the temperature change [°C], ρ is the density of the storage material [kg/m^3] and V is the volume of the storage material [m^3].

Heat loss from a sensible thermal storage is directly proportional to the temperature difference between the storage and the environment. An important consideration in sensible thermal storage systems is the rate at which heat can be released and extracted, which is a function of thermal diffusivity. Thermal conductivity, which is material property of the thermal storage, affects charging and discharging rates of the storage. This relationship is expressed by the following equation [18].

$$\lambda = \rho C_p \alpha \tag{2}$$

where λ is the thermal conductivity [W/m K], ρ is the density [kg/m^3], Cp is the specific heat [J/kg K] and α is the thermal diffusivity [m^2/s].

For a thermal energy storage system to be effective, certain requirements must be fulfilled. Requirements for the common sensible heat storage materials are high energy density (high density and specific heat) and good thermal conductivity (for residential applications usually above 0.3 W/m K). The ability of storing heat in a given container depends on the value of the quantity ρCp, the thermal capacity [19]. Thermal capacities of different storage materials are given in **Tables 1** and **2**. Most common sensible storage media include rocks, sand, pebbles packed in an insulated container. These materials have several advantages including non-toxicity, non-flammability and lower price. Thermal storage materials must be cheap and have good thermal capacity. **Table 3** lists low cost thermal storage materials, the cost ranging from 0.05 to 5.00 $/kg. The only drawback of these materials is their low heat capacities, ranging from 0.56 to 1.3 kJ/(kg °C), which can make the storage unit unrealistically large [20]. For example, one of the drawbacks of rock, sand and

Material	Temperature [°C]	Density [kg m^{-3}]	Specific heat [J kg^{-1} K^{-1}]
Clay	20	1458	879
Brick	20	1800	837
Sandstone	20	2200	712
Wood	20	700	2390
Concrete	20	2000	880
Glass	20	2710	837
Aluminum	20	2710	896
Iron	20	7900	452
Steel	20	7840	465
Gravelly earth	20	2050	1840
Magnetite	20	5177	752
Sand	20	1555	800
Rock	20	2560	879

Table 1.
Thermal capacities of selected solid storage materials [19, 20].

Material	Temperature [°C]	Fluid type	Density [kg m^{-3}]	Specific heat [J kg^{-1} K^{-1}]
Calorie HT43	12-260	Oil	867	2200
Engine oil	≤ 160	Oil	888	1880
Ethanol	≤ 78	Organic liquid	790	2400
Propane	≤ 97	Organic liquid	800	2500
Butane	≤ 118	Organic liquid	809	2400
Isotunaol	≤ 100	Organic liquid	808	3000
Isopentanol	≤ 148	Organic liquid	831	2200
Octane	≤ 126	Organic liquid	704	2400

Table 2.
Thermal capacities of selected liquid storage materials [20].

Material	Working temperature [°C]	Density [kg m-3]	Specific heat [J/ kg °C]
Sand-rock minerals	200-300	1700	1.30
Reinforced concrete	200-400	2200	0.85
Cast iron	200-400	7200	0.56
NaCl	200-500	2160	0.85
Cast steel	200-700	7800	0.60
Silica fire bricks	200-700	1820	1.00
Magnesia fire bricks	200-1200	3000	1.15

Table 3.
Solid-state low-cost sensible thermal storage materials [20].

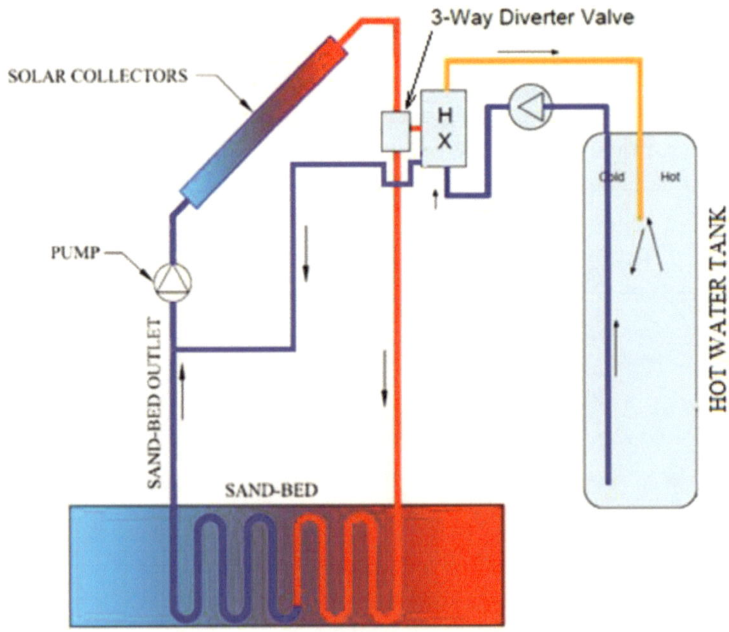

Figure 3.
Schematic of the evacuated tube solar collectors and the thermal storage.

pebbles is that larger amount is needed because of their lower heat storing capacity. The cost of the storage media per unit energy stored is, however, still acceptable for rocks [18].

2.1.1 Sensible solid heat storage system

Sensible solid heat storage media have advantages over liquids because of their higher temperature change. That is, solid thermal storages have the advantage of allowing higher temperature changes as compared to liquids. It should be noted that in sensible thermal storage, there is no phase change of the storing medium be it liquid or solid. Sensible solid storage media do not melt, therefore do not flow, hence no leakage is expected from the storage container. **Figure 3** illustrates the use of solid sensible thermal storage. The schematic shows a sand-bed thermal storage underneath a garage floor. Underneath the garage slab is the solar thermal storage which contains fine sand and pit run gravel as a thermal storage medium. The sand bed was bordered underneath with 20 cm (8″) of polystyrene foam resulting in a thermal resistivity of RSI-5.64 (US R-32) insulation barrier between the sand-bed and ground. The four sides of the sand-bed were insulated with 0.2 m (8″) of polystyrene foam board on both sides of a 0.2 m (8″) poured concrete foundation wall for a total of 0.4 m (16″) of insulating foam. Solar collectors heat a water-glycol solution that, during normal operation, passes through a heat exchanger to heat domestic hot water tank. When the domestic hot water tank is not calling for heat, the excess heat is sent to the sand bed (thermal storage) under the garage floor for heating [21]. The system has dual purpose: heating the garage by radiation and convection and heating domestic water.

An approximate rule of thumb for sizing is to use 300–500 kg of rock per square meter of collector area for space heating applications [18]. Rock or pebble-bed storages can also be used for much higher temperatures up to 1000°C [18].

2.1.2 Sensible liquid heat storage system

For liquid heat storage systems, the temperature range that can be reached is limited by their boiling points. The type of liquid used as a storage medium is determined by the desired storage temperature. Water, with its high specific heat, is the most common storage medium below 100°C temperature. Cost-effective, large-scale thermal storage is possible by using naturally occurring confined underground water such as aquifers. Hot water is pumped to be stored into such aquifers, thereby displacing the existing cold ground water. This would lower the cost of the thermal storage as the only investment required is the cost of drilling openings for injecting and withdrawing water.

If water is used for higher temperature applications, (temperatures above 100°C), it must be pressurized, adding to the cost; for such a case the limitation of water is the critical point, that is, 374°C [17]. Organic liquids and high molecular weight oils are also effective at higher temperature. Although there are oils in the market, such as Terminol, which can be used without pressurization in the range of −10 to 320°C, they have the disadvantage of being of low specific heat (2.3 KJ/kg K vs. 4.19 kJ/kg K for water). In addition, oils are liable to high-temperature cracking, polymerization and formation of volatile products. The advantages and disadvantages of water as a storage medium are listed below [18].

Advantages:

i. Relatively inexpensive, easily available, nontoxic and noncombustible,

ii. Water has a comparatively high specific heat and high density,

iii. Heat exchangers may be avoided if water is used as the heat carrier in the collector,

iv. Simultaneous charging and discharging of the storage tank are possible and

v. Adjustment and control of a water system are variable and flexible.

Disadvantages:

i. Freezing or boiling,

ii. Corrosive properties and

iii. Working temperatures are limited to less than 100°C.

Eq. (3) can be used to predict water storage temperature as a function of time [20].

$$t_s = t_i + \frac{\Delta\tau}{mC_p}\left(Q_u - Q_L - U_s A_s \left(t_f - t_a\right)\right) \tag{3}$$

where Q_u is rate of energy addition to the thermal collector, Q_L are rate of removal of energy from the collector, U_s is the heat loss coefficient of storage tank, A_s is the storage tank surface area, t_f is the final temperature, t_a is the ambient temperature for the tank and τ is the time.

Figure 4 shows schematic of a typical water tank thermal storage system. In this system, a solar thermal collector supplies the input heat and a load is served by circulating hot water through the heat exchanger. In the schematic shown, the system can also be applied for domestic hot water systems since the heat exchanger prevents contamination of potable water in domestic hot water systems.

Hot water storage systems used as buffer storage for domestic hot water supply are usually in the range of 500 L to several cubic meters. This technology is also used in solar thermal installations for domestic hot water combined with building heating systems [20].

2.1.3 Sensible cold storage system

In sensible cold storage systems, heat is removed from the storage medium. This has the effect of lowering the temperature of the storage medium. Cold storage has the potential to save operating costs. This is possible using cheaper electricity rate during off peak hours. A cold storage may consist of cold rocks or chilled water. An air-conditioning system can befit from heat sinks that can be used as cold storage to which heat is dumped. Coupling chillers to cold storages is a more efficient way of using them, although the initial investment cost is higher as compared to conventional air conditioning systems without cold storage. If water is used as cold storage medium, a large quantity is needed as its useful temperature is somewhat limited as compared to when it is used as sensible hot storage.

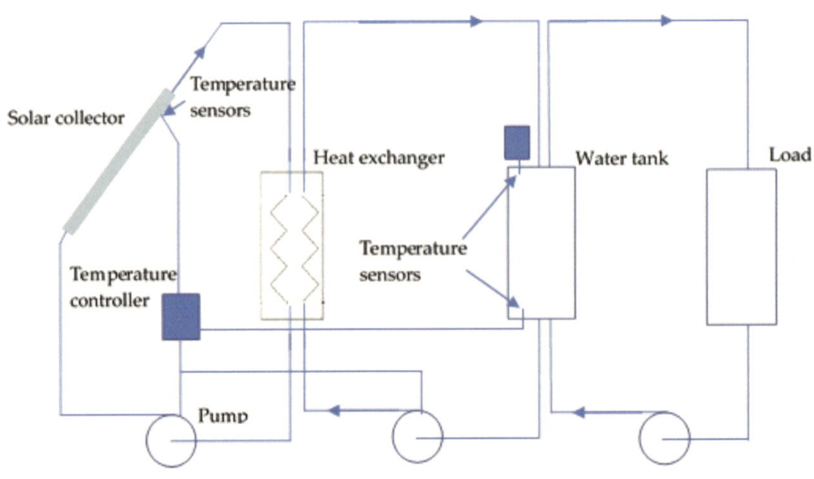

Figure 4.
Schematic of a typical water thermal storage tank.

2.2 Latent heat storage system

Latent heat storage (LHS) is based on the heat absorption or release when a storage material undergoes a phase change from solid to liquid or liquid to gas or vice versa. Latent heat thermal storage system involves the storage of energy in phase change materials (PCMs). For example, when solid material melts and turns to liquid, it absorbs heat without changing its temperature. Thermal energy is stored and released with changes in the material's phase. Latent heat storage has the advantage of being compact, that is, for a given amount of heat storage, the volume of PCM is significantly less than the volume of sensible heat storage. This leads to use of less insulation material and applicability in places where space availability is a concern. Another advantage of phase change materials is that they can be applied where there is strict working temperature as the storage can work under isothermal conditions. Latent heat storage systems also have the advantage of having high storage density. Furthermore, small temperature changes in LHS result in storing large amount of heat.

A comparison of the LHS and SHS system shows that with LHS systems, about 5–10 times higher storage densities can be obtained [18]. The volume of PCM storage is two times lesser than that of water. LHS systems can also be used over a wide temperature range [18]. Phase change can be from solid to solid, from solid to liquid, from solid to gas, from liquid to gas and vice versa. When a phase change is from solid to solid, heat is stored as the material transitions from one crystalline arrangement to another. Solid-to-solid transitions have low latent heat. Solid-to-gas and liquid-to-gas transformations are associated with higher latent heat release and higher volume change; however, the large change in volume is an issue as a huge container is required making the system more complex and impractical. As a result, the most advantageous phase change is the transition from solid to liquid (small change in volume), although solid to liquid transitions have low latent heat as compared to liquid to gas. Solid-to-liquid phase change materials are cost-effective as thermal energy storage media.

Since PCMs themselves cannot be used as heat transfer media, a separate heat transfer media must be used with heat exchanger in between to transfer energy from the source to the PCM and from PCM to the load. The heat exchanger to be used must be designed specially, in view of the low thermal diffusivity of PCMs in general. Generally, it is required that the PCM container must be compatible with the PCM and be able to handle volume changes.

The storage capacity of the LHS system with a PCM medium is given by [22]:

$$Q = m\left(C_{sp}(T_m - T_i) + a_m \Delta h_m + C_{lp}\left(T_f - T_m\right)\right) \tag{4}$$

where a_m is fraction melted, C_{lp} is average specific heat between Tm and Tf [J/kg K], C_{sp} is average specific heat between Ti and Tm [kJ/kg K], Δh_m is heat of fusion per unit mass [J/kg], m is mass of heat storage medium [kg], Q is amount of heat stored [J], T_f is final temperature [°C], T_i is initial temperature [°C] and T_m is melting temperature [°C].

When a phase change material is heated, initially it behaves like sensible heat energy storage and there is a change in temperature. That is, at the beginning, the temperature of the PCM starts to rise (**Figure 5**). Once the phase change transition temperature is reached, the PCM continues to absorb heat without changing its temperature. The PCM starts to melt and transition from solid to liquid phase. The heat absorbed at constant temperature is called the latent heat of the phase transition. **Figure 5** shows the transition process from solid liquid. It is seen from **Figure 5** that the phase transformation process occurs at a constant temperature, and the amount of heat required to carry out the process is known as latent heat. Phase-transition enthalpy of PCMs is usually much higher (100–200 times) than sensible heat,

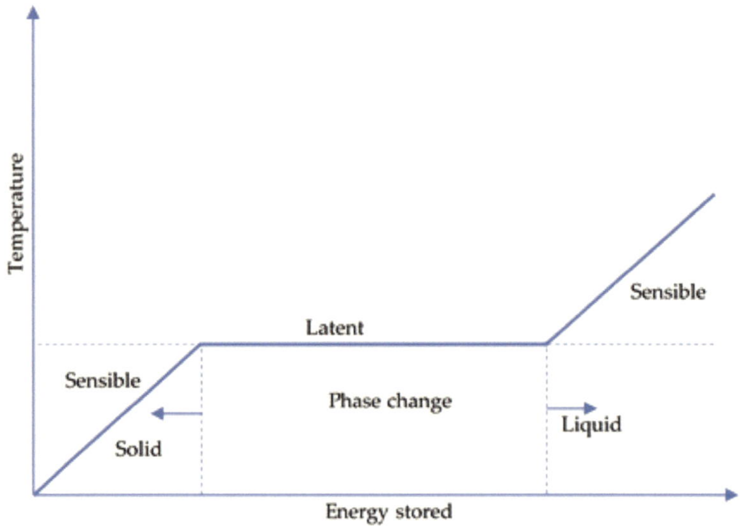

Figure 5.
Latent energy storage for solid-liquid phase change.

Storage material	Phase change temperature [°C]	Density [kg/m³]	Latent heat [kJ/kg K]	Latent heat [MJ/m³]
RT100 (paraffin)	100	880	124	N/A
RT110 (paraffin)	112	N/A	213	N/A
E117 (inorganic	117	1450	169	245
A164 (organic)	164	1500	306	459
$NaNO_3$	307	2260	172	389
KNO_3	333	2110	226	477
KOH	380	2044	149.7	306
$AlSi_{12}$	576	2700	560	1512
MgCl2	714	2140	452	967
NaCl	800	2160	492	1063
LiF	850	N/A	N/A	1800
Na_2CO_3	854	2533	275.7	698
K_2CO_3	897	2290	235.8	540
KNO_3-$NaNO_2$-$NaNO_3$	141	N/A	275	N/A
$LiNO_3$-$NaNO_3$	195	N/A	252	N/A

Table 4.
Properties of some commercial PCM materials: organic, inorganic slats and eutectics [24, 25].

consequently latent heat storages have much higher storage density than sensible heat storages [23].

Table 4 gives thermophysical properties of available commercial PCMs: organic, inorganic salts and eutectics. No single material can have all the required properties for an ideal thermal storage medium. Therefore, in practice, available material is used, and designers try to make up for the poor physical property by an adequate system design.

2.3 Classification of PCMs

Many phase change materials are available in any required temperature range. PCMs are generally divided into three main categories: organic PCMs, inorganic

PCMs and eutectics of organic and inorganic compounds. A detailed classification of PCMs is shown in **Figure 6**. **Figure 7** shows the typical range of melting enthalpy and melting temperature of common material classes used as PCM.

2.3.1 Organic PCMs

Organic PCMs have several characteristics which make them useful for latent heat storage. They are chemically stable than inorganic PCMs. They have been found to be compatible with and suitable for absorption into various building materials. One of the drawbacks of organic compounds is their initial cost, which is higher than inorganic PCMs [27]. However, the installed cost of organic PCMs is

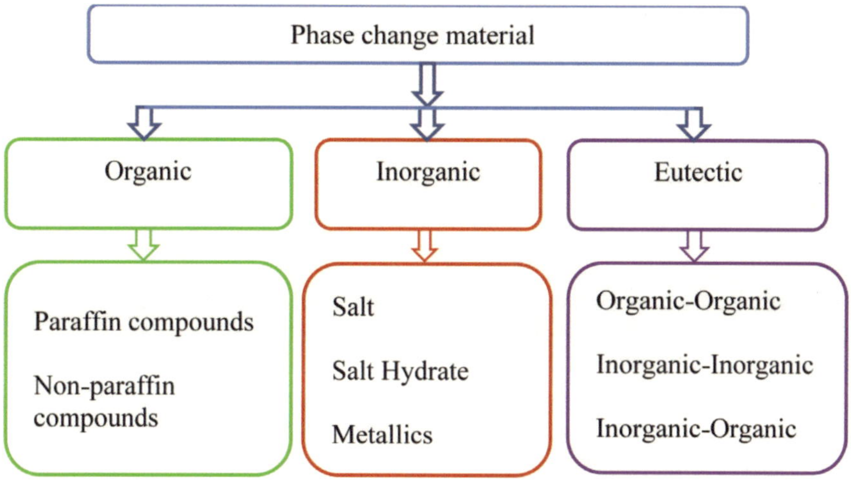

Figure 6.
Classification of PCMs [22].

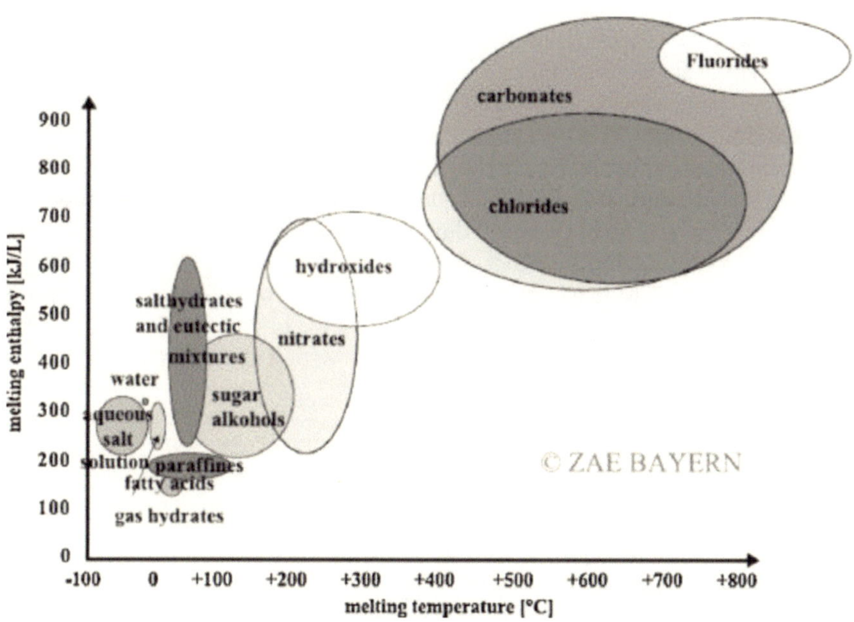

Figure 7.
Classes of materials that can be used as PCM and their typical range of melting temperature and melting enthalpy [26].

	Organic	Inorganic
Advantages	No corrosion	Greater phase change enthalpy
	Low or non subcooling	
	Chemical and thermal stability	
Disadvantages	Lower phase change enthalpy	Subcooling
	Low thermal conductivity	Corrosion
	Flammability	Phase separation
		Phase segregation
		Lack of thermal stability

Table 5.
Comparison of organic and inorganic materials for heat storage [26, 28].

competitive to inorganic PCMs. Organic PCMs are flammable, and they may generate harmful fumes on combustion, which is undesirable. They have also been found to react with the products of hydration in concrete. **Table 5** summarizes the advantages and disadvantages of organic and inorganic PCMs.

2.3.2 Inorganic PCMs

Inorganic PCMs are further classified as salts, salt hydrates and metallics. In general, inorganic PCMs have double the heat storage capacity per unit volume as compared to organic PCMs. They have a higher thermal conductivity, higher operating temperature and lower cost relative to organic phase change materials [27]. The advantages of these materials are: high latent heat values, inflammable, low-cost and availability. However, inorganic PCMs are corrosive, resulting in a short service life of the system and a higher cost [29]. Salts and salt hydrates can suffer from phase segregation and supercooling, which will reversibly affect the energy storage capacity [30]. The high storage density of salt hydrate materials is difficult to maintain and usually decreases with cycling. On the other hand, metals and metallic alloys do not suffer from phase segregation and supercooling, thus they have the potential for high temperature applications [31]. Supercooling (also called subcooling) is the process of chilling a liquid below its freezing point, without the liquid becoming a solid. This means that a temperature significantly below the melting temperature must be reached until the PCM begins to solidify and release heat. If that temperature is not reached, the PCM will not solidify at all and thus only store sensible heat [26]. **Figure 8** illustrates the process of heating/melting and cooling/solidification including subcooling.

The best-known PCM is water and it has been used for cold storage for more than 2000 years [26]. Currently, cold storage with ice is state-of-the-art. For temperatures below 0°C, usually, water-salt solutions with a eutectic composition are used [26].

2.3.3 Eutectic PCMs

Eutectic compositions are mixtures of two or more components that solidify simultaneously out of the liquid at a minimum freezing point [26]. Therefore, none of the phases can sink down due to a different density. Eutectic compositions do not segregate during melting and freezing process because they freeze to a friendly blend of crystals [32]. Supercooling is observed with many eutectic solutions [33]. Also, some eutectic solutions may be susceptible to microbiological attack; therefore, they

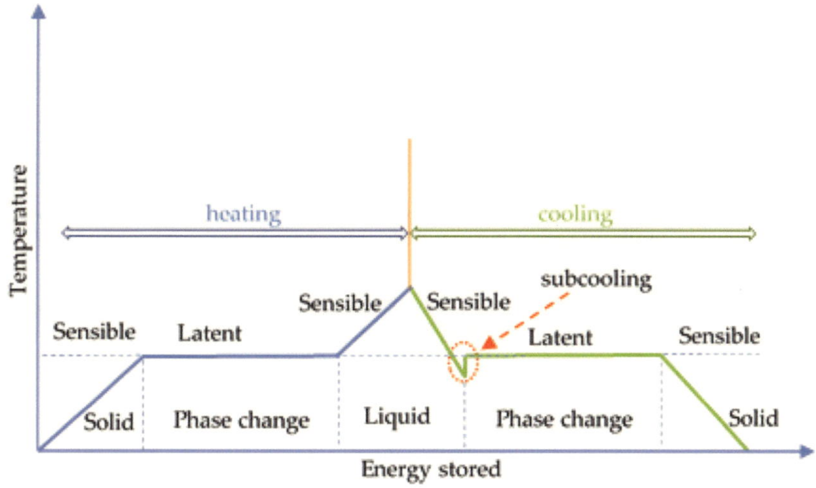

Figure 8.
Schematic of temperature change during melting and solidification of a PCM with subcooling (supercooling).

must be protected with biocides [34]. It has been reported that eutectic mixtures of fatty acid-alcohol have suitable phase change temperature, high latent heat, lower price and the potential as thermal storage materials for building energy storage [35].

2.4 Selection of appropriate PCM

The PCM to be used in the design of thermal storage systems should poses desirable thermophysical, kinetics and chemical properties [22]. As far as thermal properties are concerned, when selecting a PCM for a given application, the operating temperature of the heating or cooling should be matched to the transition temperature of the PCM. The latent heat should be as high as possible, especially on a volumetric basis, to minimize the physical size of the heat store. The following thermal properties need to be considered.

i. Suitable phase-transition temperature,

ii. High latent heat of transition and

iii. Good heat transfer.

Phase stability during freezing/melting helps in setting heat storage. High density is desirable to achieve a smaller storage size. Small volume changes during phase transformation and small vapor pressure at operating temperatures help reduce containment issues. For these reasons, the following physical properties need to be considered:

i. Favorable phase equilibrium,

ii. High density,

iii. Small volume change and

iv. Low vapor pressure.

Supercooling has been an issue with PCM development, particularly for salt hydrates. Supercooling of more than a few degrees will interfere with proper heat extraction from the storage, and 5–10°C supercooling can prevent it entirely. Thus, the following kinetic properties need to be considered in selecting a PCM:

i. No supercooling and

ii. Sufficient crystallization rate.

PCMs can degrade, chemically decompose or they can be incompatible with materials of construction. PCMs should be non-toxic, nonflammable and non-explosive for safety. Chemical properties need to be considered are as follows:

i. Long-term chemical stability,

ii. Compatibility with materials of construction,

iii. Non-toxicity and

iv. No fire hazards.

In terms of economics, for PCM to be cost-effective, in general, large-scale availability of the phase change materials is very important.

3. Improvement of PCM thermal properties

For PCMs to be cost-effective and efficient their thermal properties need to be enhanced. The enhancement of thermal properties such as thermal conductivity, latent heat and specific heat is important for the PCM under consideration to effectively transfer the heat and store more thermal energy during phase transition.

Thermal conductivity can be enhanced by: (1) inserting stationary metallic structure, (2) adding metallic nanoparticle, (3) by adding carbon additives such as graphite, graphene and carbon nanotubes (CNTs) and (4) by encapsulating the PCM [36]. Most of the researchers who studied insertion of stationary metallic structures focused on investigating configurations, shape, size and number of these insertions for optimization of thermal energy storage performance [37–39]. For example, in study conducted by Sheikholeslami et al. [40], fins and nano-enhanced phase change materials (NEPCM) were used as passive techniques to accelerate solidification process. They used finite element method to find the roles of radiation parameter, fin length and shape factor in minimizing solidification time. They used water as PCM and CuO as nanoparticles. They reported that using platelet nanoparticles leads to the greatest performance and that solid fraction of NEPCM radiation parameter has a direct relationship with solid fraction of NEPCM. As length of the fins increases, charging rate accelerates due to improvement in conduction mode [40]. In another study, Sheikholeslami investigated the effects of inner cylinder shape, diameter of nanoparticles and nanofluid volume fraction on solidification process of CuO nanoparticles and water mixture. They reported that adding nanoparticles could promote the PCM solidification with the optimum value of diameter for accelerating solidification being 40 nm [41].

Parameters such as varying volume fractions and percent weight of nanoparticles that enhance thermal performance (heat transfer) of thermal storage have been studied [42–44]. Carbon-based additives, namely graphite, graphene and carbon nanotube (CNT) have attracted much attention and are one the most promising additives that can enhance the heat transfer of PCM [36]. It has been reported that the thermal conductivity of paraffin wax can be increased by saturating porous graphite matrices in paraffin [45, 46]. The low thermal conductivity of paraffins and fatty acids may also be enhanced by using thin encapsulation, maximizing the heat transfer area [47]. Encapsulation is a method of covering the PCM (that forms the core part of the encapsulated PCM) with an appropriate coating or shell material [36]. The purpose of encapsulation is to hold the liquid and/or solid phase of the PCM and keep it isolated from the surrounding, ensuring the correct composition of the PCM that would have otherwise changed due to the mixing of the PCM with the surrounding fluid [36]. Encapsulation has also the benefit of reducing the reaction of PCM with the surrounding, providing flexibility in frequent phase change processes, increasing mechanical stability of the PCM, improving the compatibility of hazardous PCMs that cannot be directly used or immersed in certain applications such as building cooling/heating systems [36]. Based on size, encapsulated PCMs can be classified as follows [48]:

- macro (above 1 mm),
- micro (0–1000 mm) and
- nano (0–1000 nm) encapsulated PCM.

Specific heat can be enhanced by improving crystallinity of the PCM [49, 50]. A number of studies have shown that the enhanced crystallinity of PCMs in certain composites can increase specific heat capacity of the PCM [51]. In the study conducted by [49], 26 nm SiO_2 particles were added at 1% weight into a eutectic of alkali chloride salts ($BaCl_2$, NaCl, $CaCl_2$ and LiCl) with a melting point of 378°C. Addition of SiO_2 resulted in an increase of the specific heat capacity of the PCM by an over 14%, as repeatedly shown by differential scanning calorimetry (DSC) measurements. Shin et al. [50] studied further the effect in detail using SiO_2 particles having 2–20 nm diameter embedded in molten salt eutectic of Li_2CO_3-K_2CO_3. DSC measurements showed 38–54% and 118–124% increase in specific heat for the solid and the liquid phase of these composites, respectively.

Energy storage using PCM is directly dependent on the latent heat of the material. Therefore, it is always of great interest to develop materials with higher latent heat capacity. This allows either the storage of more energy within the same material mass or the use of reduced levels for a constant energy storage need [51]. Latent heat can be enhanced by increasing the crystallinity of the PCM. Warzoha and Fleischer studied the increase in latent heat resulting from the addition of multi-wall carbon nanotubes (MWCNT), alumina or TiO_2 to a base paraffin at levels of 20 vol. % concentration [52]. They found that the thermal energy that can be harnessed is 15–17% lower than the amount that can be extracted from the base paraffin during solidification; however, the thermal energy harnessed in the presence of graphene nanoparticles (15 nm thick, 15 μm diameter at 20 vol. %) is 11% higher than that for the base paraffin.

4. Challenges, opportunities and supporting polices

There are several reasons for increased deployment of energy storage technologies. The need to reduce greenhouse gas emission, the need to increase energy access and security, the need to replace aging energy infrastructure, and the need

Country/Region	Organization and overview	Type of support
Canada	**Ontario Ministry of Energy** The Ontario government will include storage technologies in its energy procurement process by the end of 2014. Initially, 50 MW of storage technologies will be installed to assist with the integration of intermittent renewable generation, optimize electric grid operation, and support innovation in energy storage technologies. Former standard offer feed-in-tariff procurement process for renewable generation projects (>500 kW) will be replaced with a competitive procurement model in Ontario. This new process will provide opportunities to consider systems that integrate energy storage with renewable energy generation.	Direct mandate, market evolution
European Union	**European Commission – Framework Research Program (FP7)** Co-funding (with the Intelligent Energy Europe Program) of the stoRE project, with the goal of creating a framework that will allow energy storage infrastructure to be developed in support of higher variable renewable energy resource penetrations. Target countries include Spain, Germany, Denmark, Austria, and Ireland.	International collaboration, policy framework development
Germany	**Federal government** Support of R&D in the framework of the energy research program and in the framework of the "funding initiative storage" Financing of a website presenting progress of funded projects.	Support of RD&D documentation, public information
Germany	**Federal Ministry of the Environment, Nature Conservation, and Nuclear Safety** Subsidy for small-scale energy storage projects to encourage distributed energy storage deployment to complement high small-scale PV penetration (2013).	Direct subsidy for distributed storage
Japan	**Ministry of Economy, Trade, and Industry (METI)** Government support of energy storage projects to demonstrate the ability to time-shift demand by 10% in conjunction with expanded use of renewable generation resources. METI funding up to 75% of storage system cost with a goal of driving down total cost of USD 234/kWh within the next seven years.	Support of demonstration projects, performance documentation
United States	**California Public Utilities Commission** Requiring the state's three largest utilities to invest in over 1.3 GW of new energy storage capacity by 2020. **FERCs – Orders 755 and 784** Taking proactive steps to open United States electricity markets to energy storage technologies Permitting companies other than large utilities to sell ancillary services in the electricity market Recognizing value of super-fast response technologies, including energy storage. Requires operators to compensate for frequency regulation based on the actual service provided. **Department of Energy** Global Energy Storage Database.	Direct mandate, market evolution, price distortion reduction, international collaboration

Table 6.
Examples of government actions that have positively supported energy storage technology deployment [12].

for decentralized energy production; all these are reasons for increased deployment of thermal energy storages. According to Grand View Research report, the global thermal energy storage market is expected to reach USD 12.50 billion by 2025 [53]. According to the report, increasing demand for access to efficient and cost competitive energy sources is expected to favor market growth. The expansion of thermal energy storage technologies is expected to be significant in Europe and Asia (particularly Japan) and somewhat lower (50%) in the United States. The global potential is estimated at approximately three times the European potential [54].

While there is great potential for thermal energy storages to be widely deployed, there are several obstacles that need to be overcome, of which the main two are cost and performance. Thermal storage market development and penetration varies considerably, depending on the application fields and regions. Penetration in the building sector is comparably slow in Europe where the construction of new buildings is around 1.3% per year and the renovation rate is around 1.5%; of course, the integration of thermal energy storage systems (TES) is easier during construction [54]. The estimate of the European potential is based on an implementation rate of TES systems in buildings of 5% [55]. Penetration could be higher in emerging countries with high rates of new buildings. TES potential for cogeneration and district heating is also associated with the building stock. The implementation rate of cogeneration is 10.2%, while the implementation of TES in these systems is assumed to be 15% [18]. As far as TES for power applications is concerned, a driving sector is concentrating solar power in which almost all new power plants in operation or under construction are equipped with TES systems, mostly based on molten salt [18].

Additional obstacles are related to material properties and stability, in particular for thermochemical systems. Each storage application needs a specific TES design to fit specific boundary conditions and requirements [18]. Most of such R&D efforts on TES technologies deal with materials (i.e., storage media for different temperature ranges), containers and thermal insulation development. For complex systems such as latent heat storage and chemical storages, more R&D is required in understanding system integration and process parameters as well as improving reacting materials.

A key to achieving widespread storage technology deployment is enabling compensation for the multiple services performed across the energy system [12]. Many governments have already acted in support of energy storage project development through efforts such as direct financial support of demonstration projects, comprehensive market transformations and mandates for energy storage projects (see **Table 6**) [12].

5. Conclusion

Thermal energy storage (TES) is a technology that works by soring thermal energy for later use. TES can be applied for heating, cooling, power generation and industrial processes. In the building area, TES are applied for use with single family houses, multi-user buildings, large commercial buildings and district heating. Most TES research has focused on materials, such as investigating storage media for different temperature ranges, containers and thermal insulation material development. PCM and thermochemical storage systems require further R&D work, for example, in the area of improving reacting materials, gaining better understanding of system integration and process parameters.

TES technologies face some barriers, cost being the key issue. Additional barriers are associated with material properties and stability, especially the thermochemical storage systems.

Penetration in the building sector is comparably slow in Europe where the construction of new buildings is around 1.3% per year and the renovation rate is around 1.5%; of course, the integration of TES is easier during construction [54]. Penetration could be higher in emerging countries with high rates of new buildings. TES potential for cogeneration and district heating is also associated with the building stock. The implementation rate of cogeneration is 10.2%, while the implementation of TES in these systems is assumed to be 15% [18].

Author details

Getu Hailu
Department of Mechanical Engineering, University of Alaska Anchorage, Anchorage, USA

Address all correspondence to: ghailu@alaska.edu

References

[1] Ciro Asperti. Native Americans of Arizona Knew the Power of Solar Energy. N.Y.: Staten Island Advance; 2013

[2] Chardin J. Travels in Persia [Internet]. The Argonaut Press. 1927. Available from: https://archive.org/stream/in.ernet.dli.2015.169543/2015.169543.Sir-John-Chardins-Travels-In-Persia_djvu.txt

[3] Chen CJ. Physics of Solar Energy. Hoboken, New Jersey and Canada: John Wiley & Sons; 2011

[4] Butti K, Perlin J. A golden thread: 2500 years of solar architecture and technology. Energy Policy. 1980:265

[5] Adams W. Cooking by solar heat. Scientific American. 1878:378

[6] Cuce E, Cuce PM. A comprehensive review on solar cookers. Applied Energy. 2013;**102**:1399-1421

[7] Saxena A, Pandey SP, Srivastav G. A thermodynamic review on solar box type cookers. Renewable and Sustainable Energy Reviews. 2011;**15**(6):3301-3318

[8] Knudson B. State of the Art of Solar Cooking—Early Uses of the Sun to Serve Humanity; 2002. Available at: http://images2.wikia.nocookie.net/__cb20070122015561/solarcooking/images/5/51/Sam.pdf

[9] Ramlow B, Nusz B. Solar Water Heating: A Comprehensive Guide to Solar Water and Space Heating Systems. US: New Society Publishers; 2006

[10] Kalogirou SA. Solar thermal collectors and applications. Progress in Energy and Combustion Science. 2004;**30**(3):231-295

[11] Spector J. How Does Thermal Energy Storage Reach Scale? [Internet]. 2017. Available from: https://www.greentechmedia.com/articles/read/how-does-thermal-energy-storage-reach-scale#gs.fo5WBII [Accessed: 14–05-2017]

[12] International Energy Agency. Technology Roadmap: Energy storage; 2014. Available at: https://www.iea.org/publications/freepublications/publication/TechnologyRoadmapEnergystorage.pdf

[13] Adamson KA, O'Donnell A. Thermal Storage for HVAC in Commercial buildings, District Cooling and Heating, Utility and Grid Support Applications, and High-Temperature Storage at CSP Facilities. PikeResearch, New York; 2012

[14] Gao L, Zhao J, Tang Z. A Review on borehole seasonal solar thermal energy storage. Energy Procedia. 2015;**70**:209-218

[15] Dincer I, Rosen MA. Thermal Energy Storage: Systems and Applications. 2nd ed. Wiley; 2010

[16] Duffie JA, Beckman WA. Solar Engineering of Thermal Processes. 4th ed. US and Canada: John Wiley and Sons; 2013

[17] Dincer I. On thermal energy storage systems and applications in buildings. Energy and Buildings. 2002;**34**(4):377-388

[18] Shah YT, editor. Thermal Energy: Sources, Recovery, and Applications. Boca Raton, FL, USA: CRC Press; 2018

[19] Dincer I. Evaluation and selection of energy storage systems for solar thermal applications. International Journal of Energy Research. 1999;**23**(12):1017-1028

[20] Sarbu I. A comprehensive review of thermal energy storage. Sustainability. 2018;**10**(2):191

[21] Hailu G, Hayes P, Masteller M. Seasonal solar thermal energy sand-bed storage in a region with extended freezing periods: Part I experimental investigation. Energies. 2017;**10**(11): 1873

[22] Sharma A, Tyagi VV, Chen CR, Buddhi D. Review on thermal energy storage with phase change materials and applications. Renewable and Sustainable Energy Reviews. 2009;**13**(2):318-345

[23] Tian Y, Zhao CY. A review of solar collectors and thermal energy storage in solar thermal applications. Applied Energy. 2013;**104**:538-553

[24] Gil A et al. State of the art on high temperature thermal energy storage for power generation. Part 1-Concepts, materials and modellization. Renewable and Sustainable Energy Reviews. 2010; **14**(1):31-55

[25] Pilkington Solar International GmbH. Survey of Thermal Storage for Parabolic Trough Power Plants Period of Performance: Survey of Thermal Storage for Parabolic Trough Power Plants Period of Performance. NREL; September 2000. p. 61

[26] Mehling H, Cabeza LF. Heat and Cold Storage with PCM: An Up to Date Introduction into Basics and Applications; Berlin Heidelberg: Springer; 2008

[27] Liu M, Saman W, Bruno F. Review on storage materials and thermal performance enhancement techniques for high temperature phase change thermal storage systems. Renewable and Sustainable Energy Reviews. 2012; **16**(4):2118-2132

[28] Zalba B. Review on thermal energy storage with phase change: Materials, heat transfer analysis and applications. Applied Thermal Engineering. 2003; **23**(3):251-283

[29] Chen C, Wang L, Huang Y. Morphology and thermal properties of electrospun fatty acids/polyethylene terephthalate composite fibers as novel form-stable phase change materials. Solar Energy Materials & Solar Cells. 2008;**92**(11):1382-1387

[30] Liu H, Awbi HB. Performance of phase change material boards under natural convection. Building and Environment. 2009;**44**(9):1788-1793

[31] Bruno F, Belusko M, Liu M, Tay NHS. Using Solid-Liquid Phase Change Materials (PCMs) in Thermal Energy Storage Systems. In: Cabeza LF, editors. Advances in Thermal Energy Storage Systems: Methods and Applications; Woodhead Publishing Series in Energy: Cambridge, UK: Woodhead Publishing is an imprint of Elsevier Sawston; 2015

[32] Mohamed SA et al. A review on current status and challenges of inorganic phase change materials for thermal energy storage systems. Renewable and Sustainable Energy Reviews. 2017;**70**:1072-1089

[33] Giriswamy BG. Experimental study and thermal characterization of nano composite phase change material. International Journal of Mechanical Engineering and Robotics Research. 2014;**3**(1):360-376

[34] Ure Z. Paper No:18 1. In: Phase Change Material (PCM) Based Energy Storage Materials And Global Application Examples; 2011. no. 18. pp. 1-24

[35] Huang J, Lu S, Kong X, Liu S, Li Y. Form-stable phase change materials based on eutectic mixture of tetradecanol and fatty acids for building energy storage: Preparation and

performance analysis. Materials (Basel). 2013;**6**(10):4758-4775

[36] Kant K, Shukla A, Sharma A. Advancement in phase change materials for thermal energy storage applications. Solar Energy Materials & Solar Cells. 2017;**172**:82-92

[37] Kousksou T, Mahdaoui M, Ahmed A, Msaad AA. Melting over a wavy surface in a rectangular cavity heated from below. Energy. 2014;**64**:212-219

[38] Pakrouh R, Hosseini MJ, Ranjbar AA, Bahrampoury R. A numerical method for PCM-based pin fin heat sinks optimization. Energy Conversion and Management. 2015;**103**:542-552

[39] Mustaffar A, Harvey A, Reay D. Melting of phase change material assisted by expanded metal mesh. Applied Thermal Engineering. 2015;**90**

[40] Saleem S, Sheikholeslami M, Ghasemi A, Li Z, Shafee A. Influence of CuO nanoparticles on heat transfer behavior of PCM in solidification process considering radiative source term. International Journal of Heat and Mass Transfer. 2018;**126**:1252-1264

[41] Sheikholeslami M. Numerical simulation for solidification in a LHTESS by means of nano-enhanced PCM. Journal of the Taiwan Institute of Chemical Engineers. 2018;**86**:25-41

[42] Ebrahimi A, Dadvand A. Simulation of melting of a nano-enhanced phase change material (NePCM) in a square cavity with two heat source-sink pairs. Alexandria Engineering Journal. 2015; **54**(4):1003-1017

[43] Fan LW, Zhu ZQ, Liu MJ. A similarity solution to unidirectional solidification of nano-enhanced phase change materials (NePCM) considering the mushy region effect. International Journal of Heat and Mass Transfer. 2015; **86**:478-481

[44] Seyed Sahand Sebti SAH, Mastiani M, Mirzaei H, Dadvand A, Kashani S. Numerical study of the melting of nano-enhanced phase change material in a square cavity. Journal of Zhejiang University. Science. A. 2013;**14**(5): 307-316

[45] Mills A, Farid M, Selman JR, Al-Hallaj S. Thermal conductivity enhancement of phase change materials using a graphite matrix. Applied Thermal Engineering. 2006;**26**(14–15): 1652-1661

[46] Karaipekli A, Biçer A, Sarı A, Tyagi VV. Thermal characteristics of expanded perlite/paraffin composite phase change material with enhanced thermal conductivity using carbon nanotubes. Energy Conversion and Management. 2017;**134**:373-381

[47] De Gracia A, Cabeza LF. Phase change materials and thermal energy storage for buildings. Energy and Buildings. 2015;**103**:414-419

[48] Shembekar PS, Salunkhe PB. A review on effect of phase change material encapsulation on the thermal performance of a system. Renewable and Sustainable Energy Reviews. 2012; **16**:5603-5616

[49] Shin D, Banerjee D. Enhancement of specific heat capacity of high-temperature silica-nanofluids synthesized in alkali chloride salt eutectics for solar thermal-energy storage applications. International Journal of Heat and Mass Transfer. 2011; **54**(5–6):1064-1070

[50] Shin D, Banerjee D. Enhanced specific heat capacity of nanomaterials synthesized by dispersing silica nanoparticles in eutectic mixtures. Journal of Heat Transfer. 2013;**135**(3): 32801

[51] Fleischer AS. Thermal Energy Storage Using Phase Change Materials. [electronic resource]: Fundamentals and Applications; Switzerland: Springer; 2015

[52] Warzoha RJ, Fleischer AS. Improved heat recovery from paraffin-based phase change materials due to the presence of percolating graphene networks. International Journal of Heat and Mass Transfer. 2014;**79**:314-323

[53] Thermal Energy Storage Market Size To Reach $12.50 Billion By 2025. New York: PR Newswire Association LLC; 2017

[54] Thermal Energy Storage. IRENA E17, IEA-ETSAP and IRENA© Technology-Policy Brief. 2013. Available at: https://iea-etsap.org/E-TechDS/PDF/E15_Ren_integr_FINAL_Dec2013_GSOK.pdf

[55] Arce P, Cabeza LF, Medrano M. GREA – Report: Potential of Energy & CO_2 savings due to the use of thermal energy storage. A Continental Overview – Europe, Universitat de Lleida, 2010

Chapter 3

Heat Transfer Enhancement Technique of PCMs and Its Lattice Boltzmann Modeling

Zhiguo Qu

Abstract

Phase change materials (PCMs) have several advantages for thermal energy storage due to their high energy storage density and nearly constant working temperature. Unfortunately, the low thermal conductivity of PCM impedes its efficiency of charging and discharging processes. To solve this issue, different techniques are developed to enhance the heat transfer capability of PCMs. In this chapter, the common approaches, which include the use of extended internal fins, porous matrices or metal foams, high thermal conductivity nanoparticles, and heat pipes for enhancing the heat transfer rate of PCMs, are presented in details. In addition, mathematical modeling plays a significant role in clarifying the PCM melting and solidification mechanisms and directs the experiments. As a powerful mesoscopic numerical approach, the enthalpy-based lattice Boltzmann method (LBM), which is robust to investigate the solid-liquid phase change phenomenon without iteration of source terms, is also introduced in this chapter, and its applications in latent heat thermal energy storage (LHTES) unit using different heat transfer enhancement techniques are discussed.

Keywords: phase change materials, heat transfer enhancement, nanofluid, mathematical models, lattice Boltzmann method

1. Introduction to heat transfer enhancement techniques of PCMs

The development of renewable energy such as solar energy and wind energy has attracted lots of attention during the past decades due to the gap between the increased global energy demand and the decreased amount of fossil fuel in the world. However, one of the major drawbacks for renewable energy is its territorial, time-dependent, and intermittent characteristics. Under this circumstance, the energy storage techniques play an indispensable role for achieving a continuous and reliable supply of renewable energy [1, 2]. Thermal energy storage (TES), which stores the heat in the materials and generates the electricity with heat engine cycles later, is a promising energy storage technique. In general, TES could be classified into three different categories, namely latent, sensible, and thermochemical. With the advantages of high energy storage density and nearly constant charging/discharging temperature, latent heat thermal energy storage (LHTES) using phase change materials (PCMs) is widely used in several renewable energy applications. However, a major issue of LHTES system is the low thermal conductivity of most

PCMs, which seriously impedes the energy storage efficiency. To handle this challenge, several heat transfer enhancement techniques are developed and discussed by researches during the past years. The existing effective approaches to ameliorate the thermal performance of PCMs include using extended internal fins, filling porous matrix or metal foams, adding high thermal conductivity nanoparticles, and using heat pipes [3].

With the characteristics of simple fabrication, low cost construction, and large heat transfer surfaces, fins are used in a majority of PCM-based LHTES systems. There are several different configurations of fins such as longitudinal, annular, circular, plates, pins, tree shape, and other novel geometries as shown in **Figure 1** [4]. By applying extended internal fins, the average thermal conductivity and heat transfer depth of LHTES system are improved, so that the melting and solidification rate of PCMs are accelerated. However, there exists a tradeoff between the increased PCM charging/discharging rate in LHTES unit and its corresponding reduced energy storage capacity with the existence of internal fins. An optimum design of fin configuration and arrangement becomes significant for achieving high energy storage efficiency of LHTES unit with PCMs. Under this circumstance, lots of researches are numerically and experimentally carried out to investigate the conjugate heat transfer between fins and PCMs, and the enhancement of PCM thermal performance with different type of fins is deeply understood and optimized during the recent years [5–21].

Due to the high surface area to volume ratio generated by the tortuosity of metal foams as shown in **Figure 2**, the PCM charging/discharging rates could be highly improved by inserting metal foams into the LHTES unit [22]. As air inevitably exists

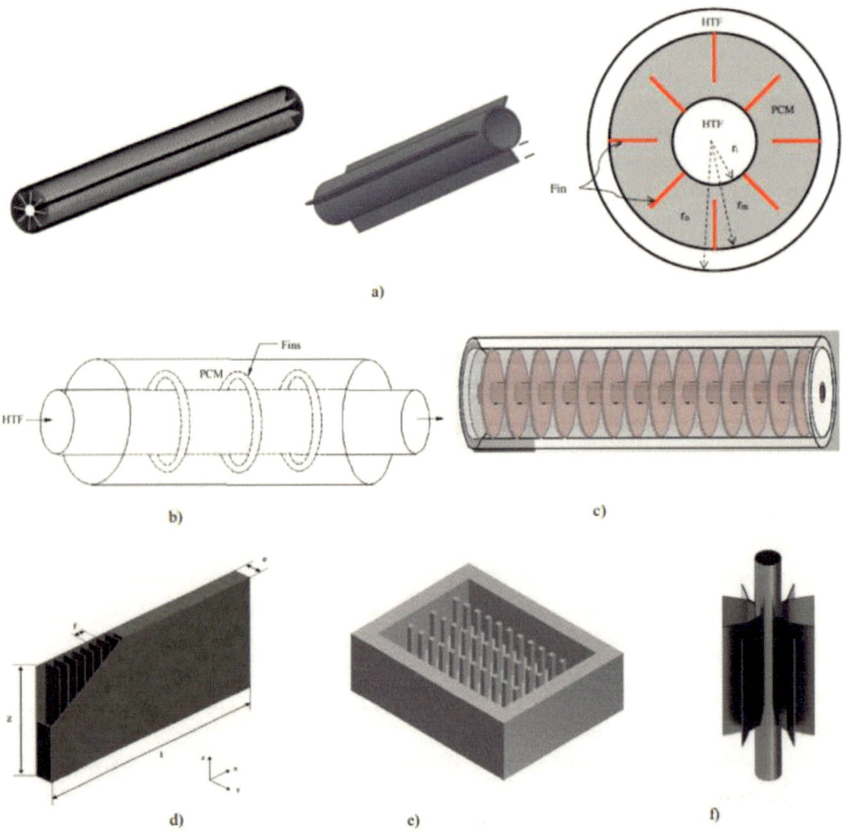

Figure 1.
Different configuration of fins used in LHTES system with PCMs [4].

Figure 2.
A piece of FeCrAlY foam sample manufactured with the sintering route [22].

in the porous structure of metal foams, the infiltration of PCMs into metal foams is hindered, which correspondingly reduces the impregnation ratio of PCMs. As a consequence, the energy storage capacity of LHTES system using combination of PCMs and metal foams is affected. To handle this difficulty, a vacuum impregnation method is generally used to prepare the PCM/metal foam composite materials. **Figure 3** shows the apparatus and procedures of impregnation treatment for PCMs with vacuum assistance, and the detailed steps could be found in Ref. [23]. As the PCM melting and solidification are actually accelerated due to the interconnected heat transfer channel inside the metal foams, the porosity and pore size of metal foams are the most significant factors, which affect the energy storage efficiency of LHTES. The copper foam with different porosities and pore sizes is displayed in **Figure 4** as an example [24]. The conduction heat transfer in the LHTES system could be consolidated with the decrement of porosity and pore sizes because of the increased density of high speed heat transfer channels inside the metal foams. However, natural convective heat transfer of liquid PCMs through the metal foams is hampered due to the reduced void space caused by the decreased metal foam porosity and pore size. In addition, when the porosity of metal foams decreases, the amount of pure PCMs in the LHTES unit is reduced, which decreases the energy storage capacity. Due to the above reasons, metal foams with appropriate porosity, pore size, and filling ratio, which balance the conduction and natural convection, are essential for achieving the optimum heat transfer rate of PCMs and the most efficient energy storage of LHTES unit. Hence, the mechanisms of PCM melting and solidification processes inserted with various metal foams are studied by several researches at both macroscopic and pore scales [25–44]. Besides, the heat transfer rate of PCMs could also be ameliorated by applying other additives such as graphite [45, 46], carbon nanotubes [47], and graphene [48].

As the nanotechnology has achieved rapid development during the past decades, adding high thermal conductivity nanoparticles becomes a new technique to improve the low heat transfer rate of PCMs [49]. Khodadadi and Hosseinizadeh first investigated the enhancement of PCM heat transfer capability using nanoparticles [50], and their results demonstrated that nanoparticle-enhanced PCMs (NEPCMs) have a great potential in TES applications. The SEM micrographs of nanoparticle-enhanced PCM ($NaNO_3$-KNO_3) with different nanoparticles and mass fractions are shown in **Figure 5** [51]. With the existence of nanoparticles, the thermophysical properties of PCMs such as thermal conductivity and latent heat capacity are varied. The mechanism of the effects of surface, chemical, and physical

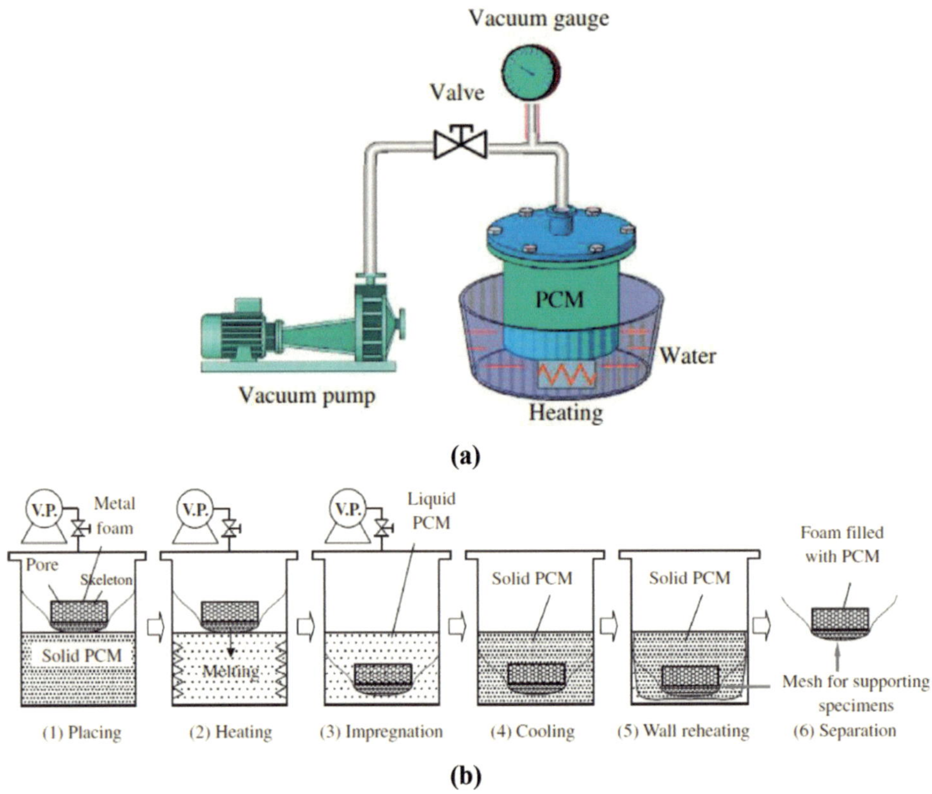

Figure 3.
Schematic of the apparatus and procedure for the preparation of composite PCMs using a vacuum impregnation method [23].

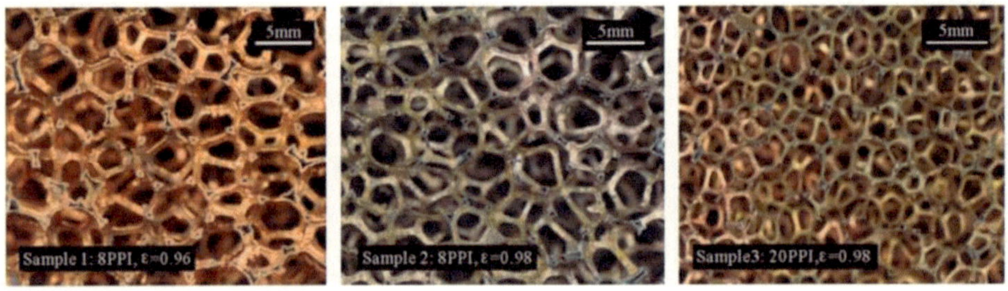

Figure 4.
Copper foam samples with different pore sizes and porosities [24].

properties of nanoparticles on the thermal properties of PCMs is reviewed in Ref. [52]. In the recent years, many researches related to NEPCMs are carried out, which mainly focus on enhancing the charging/discharging speed of PCMs with nanoparticles [53–73]. However, although the effective thermal conductivity of PCMs is ameliorated by adding nanoparticles, the energy storage capacity of LHTES unit is decreased. Furthermore, the use of nanoparticles increases the viscosity of PCMs, which impedes the development of natural convective heat transfer. Under this circumstance, the total heat transfer rate of PCMs may decrease especially for the cases under high temperature with dominant convective heat transfer. Compared with melting/solidification rate of PCMs, the energy storage rate of LHTES system is the essential goal of storing heat using PCMs. Hence, more investigations, which concentrate on the energy storage rate of NEPCM, should be completed in

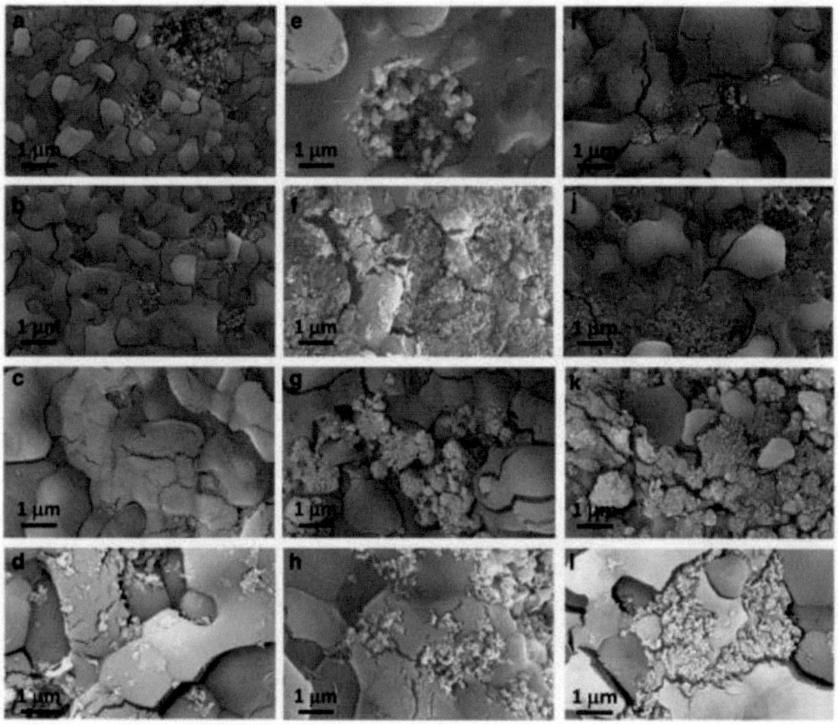

Figure 5.
SEM photos of nanoparticle-enhanced $NaNO_3$-KNO_3 with 0.5 wt.% (a–d), 1.0 wt% (e–h), and 1.5 wt% (i–l) of nanoparticles. Silica (a, e, i), alumina (b, f, j), titania (c, g, k), and alumina-silica nanoparticles (d, h, l) [51].

the future research, and the technique of adding nanoparticles for enhancing the thermal performance of LHTES needs to be further compared with other heat transfer improvement technologies in order to realize the optimum energy storage efficiency.

As the most commonly used heat exchanger devices, heat pipes are widely used to amplify the charging/discharging processes of PCMs and to transfer heat from a source to the storage or from the storage to a sink with heat transfer fluid (HTF) [74]. Although increasing the heat transfer area on the PCM side using extended fins or metal foams is the most efficient and simple method to ameliorate the energy storage rate of LHTES system as previously discussed, when there exists high temperature HTF passing through the LHTES tank such as the waste heat recovery, heat pipes are indispensable for achieving the high efficiency energy storage. The transient charging/discharging processes of PCMs in a LHTES unit with heat pipes are shown in **Figure 6** [75]. The configuration and arrangement of heat pipes play a significant role in the energy storage rate of PCMs. To optimize the thermal performance of heat pipe–assisted LHTES systems, lots of experimental and numerical works are carried out during the past few years [76–96].

The significant research progress of PCM heat transfer enhancement using a single technique is discussed and summarized in the above paragraphs. Recently, to further improve the heat transfer capability of PCM and compare the effectiveness of different approaches (use of fins, metal foams, nanoparticles, or heat pipes), the charging/discharging processes of PCMs with hybrid heat transfer enhancement techniques are investigated. Although adding nanoparticles could ameliorate the effective thermal conductivity of PCMs, the heat transfer area on the PCM side is not improved. Based on this, the extended fins are considered to be used for

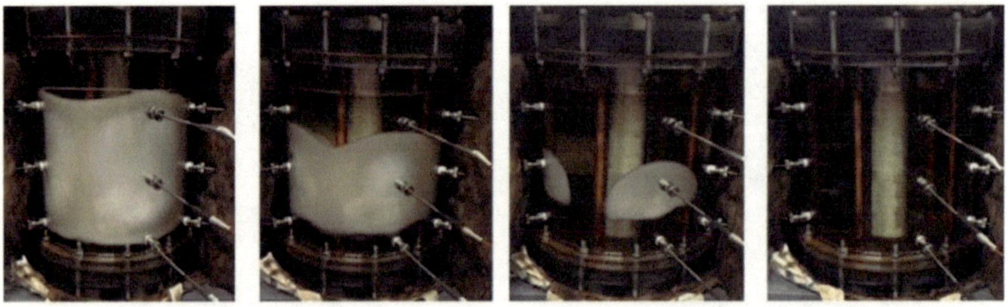

(a)

(b)

Figure 6.
The transient charging/discharging processes of PCM in a LHTES tank with HTF passing through heat pipes [75]. (a) Charging process and (b) discharging process.

enhancing the heat transfer depth of NEPCM in a LHTES unit [97–100]. Darzi et al. studied the melting and solidification of PCM enhanced with radial fins and nanoparticles in cylindrical annulus, and they found that adding fins on the hot or cold tubes is the best approach to expedite the heat transfer rate [98]. Lohrasbi et al. optimized the copper nanoparticle-PCM solidification process in a fin-assisted LHTES system [99]. The results indicate that immersing fin in LHTES unit increases the solidification rate more significantly than dispersing nanoparticles. Parsazadeh and Duan investigated the effects of fins and nanoparticle in a shell and tube LHTES unit [100]. They found that adding Al_2O_3 nanoparticles even decreases the overall heat transfer rate because the thermal conductivity enhancement with nanoparticles could not compensate for the natural convection reduction. Similarly, porous matrices are also inserted into the LHTES unit to improve the thermal performance of NEPCM [101, 102]. Hossain et al. studied the melting process of NEPCM inside the porous medium [101], and it is observed that the movement of PCM melting front is more significant under the influence of porous medium than that of nanoparticles. Tasnim et al. investigated the convection effect on the melting process of NEPCM filled in porous enclosure [102]. The results showed that both the conduction and convection heat transfer are degraded by the presence of nanoparticles. From these researches, it could be found that using extended fins or porous matrices is more effective than adding nanoparticles for enhancing the charging/discharging rate of LHTES system. Besides, other hybrid heat transfer enhancement techniques for enhancing the energy storage rate of LHTES with PCMs such as combination of fins and metal foams [103] or using combined three techniques [104–107] are also recently studied and analyzed.

In this chapter, the mathematical models for PCM charging and discharging processes with different heat transfer enhancement techniques are shown. In addition, the lattice Boltzmann method (LBM) for solid-liquid phase change phenomenon is reviewed with some classical analytical and numerical validation cases, and the implementation of graphic processor unit (GPU) computing is also presented. Furthermore, the applications of LBM modeling for LHTES system with various heat transfer improvement approaches are discussed.

2. Mathematical models

2.1 Governing equations of fluid flow and solid-liquid phase change

To simulate the charging and discharging processes of PCMs, the following assumptions are usually made to simplify the mathematical models: (1) the fluid flow of liquid PCMs is Newtonian, laminar, and incompressible with negligible viscous dissipation and (2) the thermophysical properties of PCMs, nanoparticles, fins, and metal foams are constant, except the PCM density ρ, which is a linear function of temperature T using the Boussinesq approximation. Based on the above assumptions, the flow of liquid PCMs is governed by the continuity equation and the momentum equation expressed in the Cartesian coordinate as:

$$\frac{\partial u}{\partial x}+\frac{\partial v}{\partial y}+\frac{\partial w}{\partial z}=0 \tag{1}$$

$$\rho\left(\frac{\partial u}{\partial t}+u\frac{\partial u}{\partial x}+v\frac{\partial u}{\partial y}+w\frac{\partial u}{\partial z}\right)=-\frac{\partial p}{\partial x}+\mu\left(\frac{\partial^2 u}{\partial x^2}+\frac{\partial^2 u}{\partial y^2}+\frac{\partial^2 u}{\partial z^2}\right) \tag{2}$$

$$\rho\left(\frac{\partial v}{\partial t}+u\frac{\partial v}{\partial x}+v\frac{\partial v}{\partial y}+w\frac{\partial v}{\partial z}\right)=-\frac{\partial p}{\partial y}+\mu\left(\frac{\partial^2 v}{\partial x^2}+\frac{\partial^2 v}{\partial y^2}+\frac{\partial^2 v}{\partial z^2}\right)+(\rho\beta)g(T-T_r) \tag{3}$$

$$\rho\left(\frac{\partial w}{\partial t}+u\frac{\partial w}{\partial x}+v\frac{\partial w}{\partial y}+w\frac{\partial w}{\partial z}\right)=-\frac{\partial p}{\partial z}+\mu\left(\frac{\partial^2 w}{\partial x^2}+\frac{\partial^2 w}{\partial y^2}+\frac{\partial^2 w}{\partial z^2}\right) \tag{4}$$

where ρ is the density; p is the pressure; μ is the dynamic viscosity; t is the time; β is the thermal expansion coefficient of PCMs; g is the magnitude of gravitational acceleration; T is the temperature; T_r is the reference temperature; x is the horizontal coordinate; y is the vertical coordinate; z is the coordinate, which is orthogonal with x and y coordinates; and u, v, and w are the fluid velocities in the x, y, and z directions, respectively.

The solid-liquid phase change process of PCMs is governed by the enthalpy equation as:

$$\frac{\partial(\rho H)}{\partial t}+\rho c_p\left(u\frac{\partial T}{\partial x}+v\frac{\partial T}{\partial y}+w\frac{\partial T}{\partial z}\right)=k\left(\frac{\partial^2 T}{\partial x^2}+\frac{\partial^2 T}{\partial y^2}+\frac{\partial^2 T}{\partial z^2}\right) \tag{5}$$

where c_p and k are the specific heat and thermal conductivity of PCMs. The enthalpy H of PCMs is defined as:

$$H=c_p(T-T_r)+f_l L \tag{6}$$

where f_l is the PCM liquid fraction, and L is the latent heat of PCMs. By calculating the enthalpy H of PCMs, the liquid fraction f_l and temperature T could be updated by the following equations:

$$f_l = \begin{cases} 0, & H \le H_s \\ \dfrac{H - H_s}{H_l - H_s}, & H_s < H < H_l \\ 1 & H \ge H_l \end{cases} \tag{7}$$

$$T = \begin{cases} T_m - \dfrac{H_s - H}{c_p}, & H \le H_s \\ T_m, & H_s < H < H_l \\ T_m + \dfrac{H - H_l}{c_p} & H \ge H_l \end{cases} \tag{8}$$

where H_s is the enthalpy of solid state PCMs, H_l is the enthalpy of liquid state PCMs, and T_m is the melting/solidification temperature of PCMs.

2.2 Nanofluid models

For simulating the fluid flow and heat transfer of nanoparticle-enhanced PCMs (NEPCMs), the nanoparticle is assumed to be spherical in shape, so that the Brinkman model and the Maxwell model for nanofluid are valid. In addition, the NEPCMs are considered as continuous media with the thermal dispersion being neglected. Based on this, the effective viscosity of NEPCMs is computed using the Brinkman model as [108]:

$$\mu_{nf} = \frac{\mu_{PCM}}{(1 - \Phi)^{2.5}} \tag{9}$$

where Φ is the volume fraction of nanoparticles, μ_{PCM} is the dynamic viscosity of pure PCM, and μ_{nf} is the dynamic viscosity of NEPCMs. The thermal conductivity of NEPCMs is calculated according to the Maxwell model as [109]:

$$\frac{k_{nf}}{k_{PCM}} = \frac{k_p + 2k_{PCM} - 2(k_{PCM} - k_p)\Phi}{k_p + 2k_{PCM} + (k_{PCM} - k_p)\Phi} \tag{10}$$

where k_{PCM}, k_p, and k_{nf} are the thermal conductivities of pure PCMs, nanoparticles, and NEPCMs, respectively. Furthermore, the density of nanofluid ρ_{nf} is computed using interpolation as:

$$\rho_{nf} = (1 - \Phi)\rho_{PCM} + \Phi\rho_p \tag{11}$$

where ρ_{PCM} and ρ_p are the densities of pure PCM and nanoparticles. The heat capacitance of NEPCMs $(\rho c_p)_{nf}$ is defined as:

$$(\rho c_p)_{nf} = (1 - \Phi)(\rho c_p)_{PCM} + \Phi(\rho c_p)_p \tag{12}$$

where $(\rho c_p)_{PCM}$ is the heat capacitance of pure PCM, and $(\rho c_p)_p$ is the heat capacitance of nanoparticles. The thermal expansion volume of NEPCMs $(\rho\beta)_{nf}$ is given as:

$$(\rho\beta)_{nf} = (1 - \Phi)(\rho\beta)_{PCM} + \Phi(\rho\beta)_p \tag{13}$$

where $(\rho\beta)_{PCM}$ and $(\rho\beta)_p$ are the thermal expansion volume of pure PCM and nanoparticles, respectively. The latent heat of NEPCMs $(\rho L)_{nf}$ is computed as:

$$(\rho L)_{nf} = (1 - \Phi)(\rho L)_{PCM} \tag{14}$$

where $(\rho L)_{PCM}$ is the latent heat of pure PCM. Then, the corresponding enthalpy of NEPCMs H_{nf} is given as:

$$H_{nf} = c_{p_{nf}}(T - T_r) + f_l L_{nf} \tag{15}$$

2.3 Conjugate heat transfer between PCMs and fins or metal foams

When the extended fins or metal foams are used as the heat transfer enhancement techniques for LHTES unit, the conjugate heat transfer occurs between the PCMs and the fins or metal foams. The heat transfer inside the fins or metal foams is governed by the conduction equation as:

$$\rho_s c_{p_s} \frac{\partial T}{\partial t} = k_s \left(\frac{\partial^2 T}{\partial x^2} + \frac{\partial^2 T}{\partial y^2} + \frac{\partial^2 T}{\partial z^2} \right) \tag{16}$$

ρ_s, c_{p_s}, and k_s are the density, specific heat, and thermal conductivity of fins or metal foams, respectively. On the interface of PCMs and fins or metal foams, the coupled Dirichlet-Neumann boundary conditions for conjugate heat transfer should be satisfied:

$$-k_{PCM} \frac{\partial T_{PCM}}{\partial n} = -k_s \frac{\partial T_s}{\partial n} \tag{17}$$

$$T_{PCM} = T_s \tag{18}$$

T_{PCM} and T_s are the temperature of PCMs and fins or metal foams, k_{PCM} is the thermal conductivity of PCMs, and n is the normal unit of the interface. The above boundary conditions could be satisfied in the lattice Boltzmann method automatically by using the harmonic mean value of thermophysical properties of PCMs and fins or metal foams as discussed in the following section. To investigate the melting and solidification processes of PCMs filled with metal foams at pore scale, the morphology of the real metal foam structures could be reconstructed using the

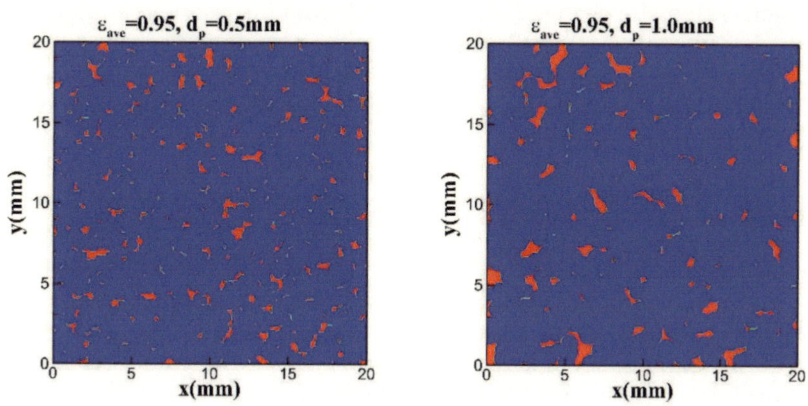

Figure 7.
Reconstructed PCM filled with metal foams using QSGS [111].

quartet structure generation set (QSGS) as shown in **Figure 7** [110, 111], where ε_{ave} and d_p are the porosity and the pore size of metal foams. Besides, the metal foams could also be reconstructed with scanning electron microscopy (SEM).

3. Lattice Boltzmann method for solid-liquid phase change

3.1 The development history of enthalpy-based LBM for solid-liquid phase change

As a mesoscopic numerical approach developed during the past more than two decades, the lattice Boltzmann method (LBM) has become a powerful tool for simulating complex heat transfer and fluid flow problems such as single-phase flows, multiphase flows, turbulence flows, flows in porous media, heat transfer, phase change, and transport in microfluidics [112–115]. The mesoscopic nature of LBM makes it appropriate for tackling the evolvement of solid-liquid interface during the phase change process. The existing LBM for solid-liquid phase change problems could be generally categorized as follows: (1) the phase-field based method, (2) the enthalpy-based method, and (3) the immersed boundary method. For the phase-field method, an auxiliary parameter, which varies smoothly across the diffusive phase interface, is used to track the solid-liquid interface implicitly [116, 117]. Unfortunately, the extremely finer grids are indispensable in the interfacial region, which increases the computational effort. In addition, Huang and Wu developed the immersed boundary-thermal lattice Boltzmann method for modeling the solid-liquid phase change phenomenon [118]. The melting interface is explicitly tracked by the Lagrangian grids, and the temperature and fluid flow are solved on the Eulerian nodes. However, an interpolation between the Lagrangian nodes and the Eulerian nodes should be carried out using the Dirac delta function, which reduces the computation accuracy. Compared with other methods, the enthalpy-based method becomes attractive for solid-liquid phase change due to its high efficiency and robustness. Jiang et al. first used LBM with enthalpy formulation to investigate the phase change problems [119]. However, the convective heat transfer is not considered in this work. Huber et al. developed a LBM model using double distribution functions to simulate the solid-liquid phase change process with convective heat transfer [120], and the numerical results are analyzed and compared with scaling laws and previous numerical work. Eshraghi and Felicelli developed an implicit LBM scheme to investigate the conduction with phase change [121]. Different from the previous enthalpy-based LBM, the iteration of nonlinear latent heat source term in the energy equation is avoided by solving a linear system of equations. Feng et al. further extended this implicit model with a consideration of natural convection to investigate the melting process of nanoparticle-enhanced PCMs [122]. To improve the computational efficiency, Huang et al. modified the equilibrium distribution function of temperature for enthalpy, so that the iteration of heat source term or solving a linear system of equations is not indispensable [123]. Unfortunately, the stability range of relaxation time for this model is narrow, which limits its applications for real scientific and engineering problems. To handle this drawback, Huang and Wu improved their work by using a multiple-relaxation-time (MRT) collision scheme instead of the single-relaxation-time (SRT) one, so that the numerical stability is highly ameliorated [124]. In addition, the thermal conductivity and the specific heat capacity are decoupled from the relaxation time and equilibrium distribution function, which make this model appropriate for satisfying the Dirichlet-Neumann boundary conditions for conjugate heat transfer. As the model developed by Huang et al. is limited in two-dimensional cases, Li et al. recently developed the SRT and MRT models for axisymmetric and three-dimensional solid-liquid phase change problems [125, 126]. Besides, to speed up the computation, Su and Davidson worked out a mesoscopic scale timestep adjustable non-dimensional LBM [127], and the time steps can be adjusted independent of mesh size by changing the transient mesoscopic Mach number. Furthermore, although the key advantage of LBM is to carry out the pore-scale numerical modeling of heat transfer in porous media, a few lattice Boltzmann models for solid-liquid phase change in porous media at the representative elementary volume (REV) scale are also developed [128–130]. The classical enthalpy-based MRT lattice Boltzmann model developed by Huang and Wu is reviewed in this section because of its simplicity and wide application by researchers for latent heat thermal energy storage problems.

3.2 Multiple-relaxation-time (MRT) method

3.2.1 MRT LBM for fluid flow

The general two dimensional nine-velocity (D2Q9) MRT LBM model is presented in this part for simulating the fluid flow. In the D2Q9 model, lattice velocities $\boldsymbol{e}_i$ are given by:

$$\boldsymbol{e}_i = \begin{cases} \boldsymbol{e}_0 = (0,0) \\ \boldsymbol{e}_i = c(\cos[(i-1)\pi/2], \sin[(i-1)\pi/2]), & i = 1,2,3,4 \\ \boldsymbol{e}_i = \sqrt{2}c(\cos[(2i-9)\pi/4], \sin[(2i-9)\pi/4]), & i = 5,6,7,8 \end{cases} \tag{19}$$

where c is the lattice speed, and the collision step carried out in the momentum space is given as:

$$\boldsymbol{m}_f(\boldsymbol{x}, t+\delta_t) = \boldsymbol{m}_f(\boldsymbol{x}, t) - \boldsymbol{S}\left[\boldsymbol{m}_f(\boldsymbol{x}, t) - \boldsymbol{m}_f^{eq}(\boldsymbol{x}, t)\right] + \delta_t\left(\boldsymbol{I} - \frac{\boldsymbol{S}}{2}\right)\boldsymbol{F}_m(\boldsymbol{x}, t) \tag{20}$$

where $\boldsymbol{x}$ is the location vector, t is the time, δ_t is the time step, $\boldsymbol{I}$ is the unit matrix, and $\boldsymbol{m}_f$ is the distribution function in momentum space defined as:

$$\boldsymbol{m}_f(\boldsymbol{x}, t) = \left[m_{f0}(\boldsymbol{x}, t), m_{f1}(\boldsymbol{x}, t), \ldots, m_{f8}(\boldsymbol{x}, t)\right]^T \tag{21}$$

The equilibrium distribution function in momentum space $\boldsymbol{m}_f^{eq}$ is [131–133]:

$$\boldsymbol{m}_f^{eq} = \left(\rho, -2\rho + 3\rho\frac{|\boldsymbol{u}|^2}{c^2}, \rho - 3\rho\frac{|\boldsymbol{u}|^2}{c^2}, \rho\frac{u_x}{c}, -\rho\frac{u_x}{c}, \rho\frac{u_y}{c}, -\rho\frac{u_y}{c}, \rho\frac{u_x^2 - u_y^2}{c^2}, \rho\frac{u_x u_y}{c^2}\right)^T \tag{22}$$

where ρ is the density, $\boldsymbol{u}$ is the fluid velocity vector, u_x is the horizontal velocity, and u_y is the vertical velocity. The diagonal relaxation matrix $\boldsymbol{S}$ is defined as:

$$\boldsymbol{S} = \mathrm{diag}(s_0, s_e, s_\varepsilon, s_j, s_q, s_j, s_q, s_p, s_p) \tag{23}$$

where s_0, s_e, s_ε, s_j, s_q, and s_p are the parameters related to relaxation time, which could be chosen as described in Ref. [134]. The discrete force term in the momentum space $\boldsymbol{F_m}(\boldsymbol{x},t)$ is given by [135, 136]:

$$\boldsymbol{F_m}(\boldsymbol{x},t)=\left(0,6\frac{\boldsymbol{F}\bullet\boldsymbol{u}}{c^2},-6\frac{\boldsymbol{F}\bullet\boldsymbol{u}}{c^2},\frac{F_x}{c},-\frac{F_x}{c},\frac{F_y}{c},-\frac{F_y}{c},2\frac{F_xu_x-F_yu_y}{c^2},\frac{F_xu_y+F_yu_x}{c^2}\right)^T \tag{24}$$

where $\boldsymbol{F}$ is the body force of fluid flow, F_x is the body force in the horizontal direction, and F_y is the body force in the vertical direction. After the collision process, the post collision distribution function in the velocity space $f_i(\boldsymbol{x},t+\delta_t)$ is calculated through inverse transformation:

$$f_i(\boldsymbol{x},t+\delta_t)=\boldsymbol{M}^{-1}\boldsymbol{m_f}(\boldsymbol{x},t+\delta_t) \tag{25}$$

where the dimensionless orthogonal transformation matrix $\boldsymbol{M}$ is chosen as [137]:

$$\boldsymbol{M}=\left\{\begin{matrix} 1 & 1 & 1 & 1 & 1 & 1 & 1 & 1 & 1 \\ -4 & -1 & -1 & -1 & -1 & 2 & 2 & 2 & 2 \\ 4 & -2 & -2 & -2 & -2 & 1 & 1 & 1 & 1 \\ 0 & 1 & 0 & -1 & 0 & 1 & -1 & -1 & 1 \\ 0 & -2 & 0 & 2 & 0 & 1 & -1 & -1 & 1 \\ 0 & 0 & 1 & 0 & -1 & 1 & 1 & -1 & -1 \\ 0 & 0 & -2 & 0 & 2 & 1 & 1 & -1 & -1 \\ 0 & 1 & -1 & 1 & -1 & 0 & 0 & 0 & 0 \\ 0 & 0 & 0 & 0 & 0 & 1 & -1 & 1 & -1 \end{matrix}\right\} \tag{26}$$

Then, the streaming step is carried out as:

$$f_i(\boldsymbol{x}+\boldsymbol{e}_i\delta_t,t+\delta_t)=f_i(\boldsymbol{x},t+\delta_t) \tag{27}$$

The nonslip velocity condition on the diffusive interface and in the solid phase is tackled by recalculating the density distribution function f_i through a linear interpolation as [138]:

$$f_i=f_lf_i+(1-f_l)f_i^{eq}(\rho,\boldsymbol{u_s}) \tag{28}$$

f_l is the liquid fraction of PCM, f_i^{eq} is the equilibrium distribution function, which could be calculated by inverse transformation as $f_i^{eq}=\boldsymbol{M}^{-1}\boldsymbol{m}_f^{eq}$. For the solid phase, there is $\boldsymbol{u_s}=\boldsymbol{0}$. Hence, the macroscopic variables, density ρ and velocity $\boldsymbol{u}$, are defined as:

$$\rho=\sum_{i=0}^{8}f_i,\ \rho\boldsymbol{u}=\sum_{i=0}^{8}\boldsymbol{e}_if_i+\frac{\delta_t}{2}\boldsymbol{F} \tag{29}$$

As mentioned in Ref. [138], the density ρ in the term $f_i^{eq}(\rho,\boldsymbol{u_s})$ in Eq. (28) should be first calculated by Eq. (29) in order to ensure the mass conservation. Then, for the liquid phase of $f_l=1$, the above lattice Boltzmann model recovers to the standard scheme for incompressible flow. On the other hand, for the solid phase of $f_l=0$, the above model could satisfy that $f_i=f_i^{eq}(\rho,\boldsymbol{u_s})$ indicating that the nonslip velocity $\boldsymbol{u}=\boldsymbol{u_s}$ is ensured.

3.2.2 MRT LBM for solid-liquid phase change

The MRT lattice Boltzmann equation (LBE) for the total enthalpy H distribution function $\boldsymbol{m}_g(\boldsymbol{x}, t)$ is expressed as [124]:

$$\boldsymbol{m}_g(\boldsymbol{x}, t+\delta_t) = \boldsymbol{m}_g(\boldsymbol{x}, t) - \boldsymbol{S}\left[\boldsymbol{m}_g(\boldsymbol{x}, t) - \boldsymbol{m}_g^{eq}(\boldsymbol{x}, t)\right] \tag{30}$$

where $\boldsymbol{m}_g$ is the distribution in momentum space given as:

$$\boldsymbol{m}_g(\boldsymbol{x}, t) = \left[m_{g0}(\boldsymbol{x}, t), m_{g1}(\boldsymbol{x}, t), \ldots, m_{g8}(\boldsymbol{x}, t)\right]^T \tag{31}$$

The equilibrium moment $\boldsymbol{m}_g^{eq}$ is given by:

$$\boldsymbol{m}^{eq} = \left(H, -4H + 2c_{p,ref}T + 3c_pT\frac{\boldsymbol{u}^2}{c^2}, 4H - 3c_{p,ref}T - 3c_pT\frac{\boldsymbol{u}^2}{c^2}, c_pT\frac{u_x}{c}, -c_pT\frac{u_x}{c}, c_pT\frac{u_y}{c}, -c_pT\frac{u_y}{c}, c_pT\frac{u_x^2 - u_y^2}{c^2}, c_pT\frac{u_xu_y}{c^2}\right)^T \tag{32}$$

where T is the temperature, and c_p is the specific heat calculated by interpolation as:

$$c_p = (1 - f_l)c_{p,s} + f_lc_{p,l} \tag{33}$$

$c_{p,s}$ is the specific heat of PCM at solid state, and $c_{p,l}$ is the specific heat of PCM at liquid state. To achieve good numerical stability, the reference specific heat $c_{p,ref}$ is defined by the harmonic mean of $c_{p,s}$ and $c_{p,l}$ as:

$$c_{p,ref} = \frac{2c_{p,s}c_{p,l}}{c_{p,s} + c_{p,l}} \tag{34}$$

The parameters in the diagonal relaxation matrix $\boldsymbol{S}$ satisfy $s_0 = 1$, $s_e = s_p$, $s_j = \frac{1}{\tau}$, and $0 < s_{e,\varepsilon,q} < 2$, where the relaxation time τ is given as:

$$\frac{k}{\rho c_{p,ref}} = c_s^2(\tau - 0.5)\delta_t \tag{35}$$

where k is the thermal conductivity, and $c_s = \frac{c}{\sqrt{3}}$ is the sound speed. To reduce the numerical diffusion, a "magic" relationship is found by Huang and Wu [124] as:

$$\left(\frac{1}{s_e} - \frac{1}{2}\right)\left(\frac{1}{s_j} - \frac{1}{2}\right) = \frac{1}{4} \tag{36}$$

Similar to the computation of fluid flow, the post-collision distribution function in the velocity space g_i could be calculated by inverse transformation as:

$$g_i(\boldsymbol{x}, t+\delta_t) = \boldsymbol{M}^{-1}\boldsymbol{m}_g(\boldsymbol{x}, t+\delta_t) \tag{37}$$

The streaming process is completed as:

$$g_i(\boldsymbol{x} + \boldsymbol{e}_i\delta_t, t+\delta_t) = g_i(\boldsymbol{x}, t+\delta_t) \tag{38}$$

Then, the enthalpy H is calculated as:

$$H = \sum_{i=0}^{8} g_i \tag{39}$$

Furthermore, it should be pointed out that the nonequilibrium extrapolation method developed by Guo et al. could be applied for the boundary conditions of fluid flow and enthalpy on the surfaces of LHTES unit [139, 140].

3.3 Classical examples for code validation

3.3.1 One-dimensional transient conjugate heat transfer

To validate the capability of MRT LBM for tackling the differences in thermophysical properties, the one-dimensional transient conjugate heat transfer in two regions without phase change is used to compare the numerical results with analytical solutions. Initially, T is equal to 1 in the region A at $x>0$, and T is equal to 0 in the region B at $x<0$. The analytical solution for this problem is given as [141]:

$$T^A(x,t) = \frac{1}{1+\sqrt{(\rho C_p)^B k^B/(\rho C_p)^A k^A}}\left[1+\sqrt{(\rho C_p)^B k^B/(\rho C_p)^A k^A}\, erf\left(\frac{x}{2\sqrt{k^A t/(\rho C_p)^A}}\right)\right] \tag{40}$$

$$T^B(x,t) = \frac{1}{1+\sqrt{(\rho C_p)^B k^B/(\rho C_p)^A k^A}}\,\mathrm{erfc}\left(-\frac{x}{2\sqrt{k^B t/(\rho C_p)^B}}\right) \tag{41}$$

When the aluminum is used for region A while the liquid water is chosen as the material in region B, the comparison between LBM results and analytical solutions is presented in **Figure 8**. A good agreement is observed, and the maximum L2 error with 200×200 lattice grids is only 6.7983×10^{-5} [142].

3.3.2 One-dimensional melting by conduction

In order to verify the MRT LBM for solid-liquid phase change phenomenon, the one-dimensional melting by conduction at a constant phase change temperature T_m is simulated. Initially, the substance is uniformly solid at a temperature T_0 ($T_0<T_m$). The melting process begins at time $t = 0$ when the temperature of the left wall is at a high temperature of T_h ($T_h > T_m$). Then, the analytical solution for the temperature in this problem is given as [143]:

$$T(x,t) = T_h - \frac{T_h - T_m}{\mathrm{erf}(\lambda)}\,\mathrm{erf}\left(\frac{x}{2\sqrt{k_l t/(\rho c_{p,l})}}\right),\ 0\leq x\leq X_i(t),\ \text{liquid} \tag{42}$$

$$T(x,t) = T_0 + \frac{T_m - T_0}{\mathrm{erfc}(\lambda/\sqrt{R_{ac}})}\,\mathrm{erfc}\left(\frac{x}{2\sqrt{k_s t/(\rho c_{p,s})}}\right),\ x>X_i(t),\ \text{solid} \tag{43}$$

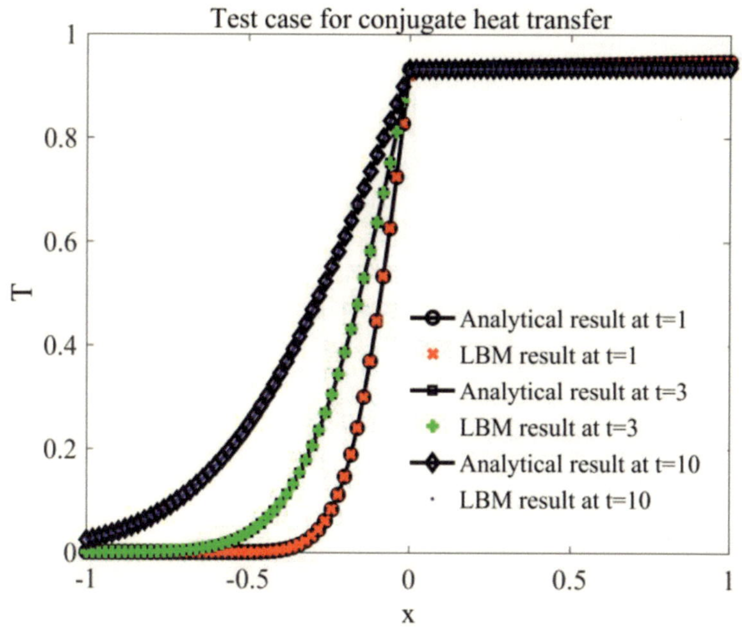

Figure 8.
One-dimensional transient conjugate heat transfer [142].

where k_l and k_s are the thermal conductivities of liquid PCM and solid PCM, respectively. The parameter $R_{ac} = \frac{[k_s t/(\rho c_{p,s})]}{[k_l t/(\rho c_{p,l})]}$ is the ratio of thermal diffusivity between the solid and liquid phases of PCM, and the location of phase interface X_i is calculated as:

$$X_i = 2\lambda\sqrt{k_l t/(\rho c_{p,l})} \quad (44)$$

The parameter λ is the root of the transcendental equation:

$$\frac{Ste_l}{\exp(\lambda^2)\mathrm{erf}(\lambda)} - \frac{Ste_s\sqrt{R_{ac}}}{\exp(\lambda^2/R_{ac})\mathrm{erfc}(\lambda/\sqrt{R_{ac}})} = \lambda\sqrt{\pi} \quad (45)$$

where the Stefan numbers, Ste_l and Ste_s, are defined as:

$$Ste_l = \frac{c_{p,l}(T_h - T_m)}{L}, \; Ste_s = \frac{c_{p,s}(T_m - T_0)}{L} \quad (46)$$

The comparison of temperature T between the analytical solutions and the SRT or MRT LBM with different relaxation times is presented in **Figure 9**. It could be observed that the numerical diffusion exists for the small (τ_l close to 0.5) and large (τ_l>2) relaxation times when the SRT model is applied. However, the numerical diffusion is highly reduced once the MRT model with a “magic” parameter relation shown in Eq. (36) is used.

3.3.3 Two-dimensional melting by convection

The natural convection with melting in a square cavity heated from the side wall is usually used to validate the code for solid-liquid phase change. First, the

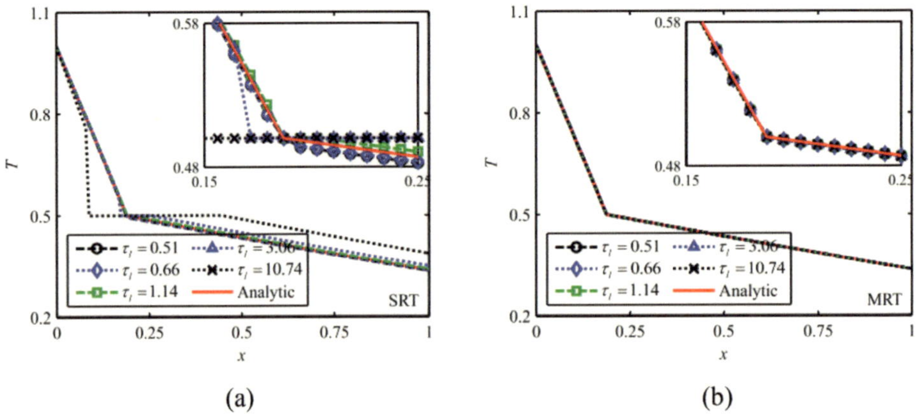

Figure 9.
(a) SRT LBM (b) MRT LBM. Comparison of temperature T between analytical solutions and SRT or MRT LBM results with different relaxation times [124].

solid-liquid phase change with convection in a cavity at the Rayleigh number $Ra = 25000$, the Prandtl number $\Pr = 0.02$, and the Stefan number $Ste = 0.01$ is compared with the results by Mencinger [144]. The average Nusselt number Nu_{ave} at the hot wall in terms of the Fourier numbers Fo is plotted in **Figure 10(a)**, and the melting interface positions at $Fo = 10$ and $Fo = 20$ are shown in **Figure 10(b)**. In addition, the average melting fraction f_l is also presented in **Figure 10(c)**. It is

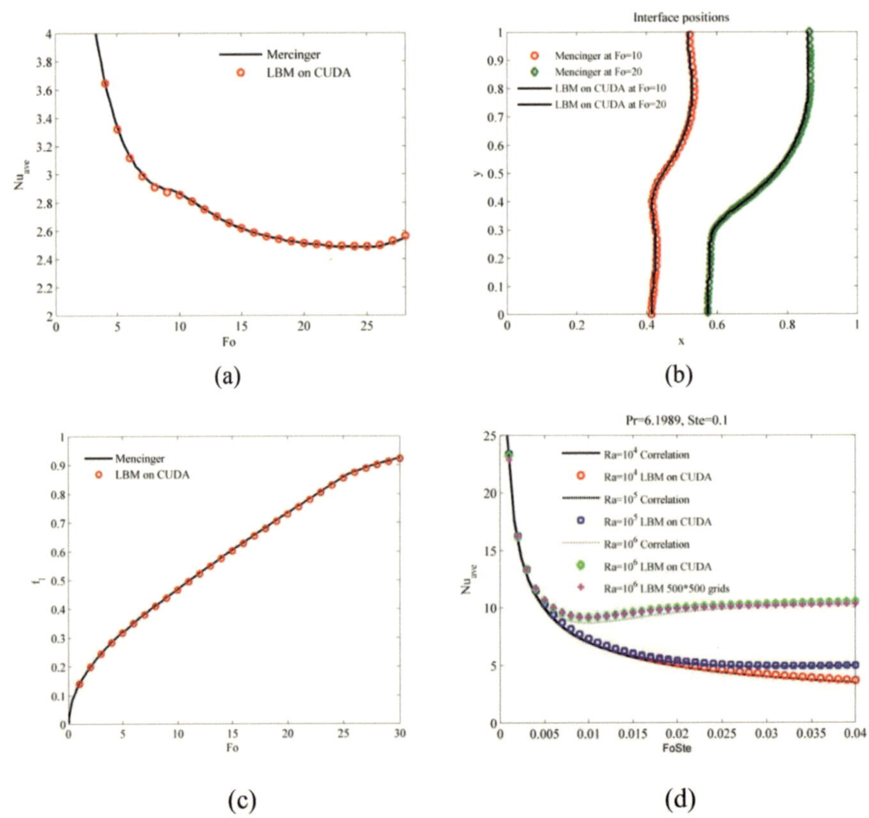

Figure 10.
Natural convection with melting in a cavity [142].

obvious that the current LBM results are consistent with the work of Mencinger [142]. Besides, for the cases of Pr>1, the MRT LBM for natural convection with melting in cavity could be calibrated by the scaling laws and correlations derived by Jany and Bejan [145]. The average Nusselt number Nu_{ave} at hot wall is given as:

$$Nu_{ave} = (2FoSte)^{-1/2} + \left[0.33Ra^{1/4} - (2FoSte)^{-1/2}\right]\left\{1 + \left[0.0175Ra^{3/4}(FoSte)^{3/2}\right]^{-2}\right\}^{-1/2} \tag{47}$$

As shown in **Figure 10(d)**, the average Nusselt number Nu_{ave} at $\mathrm{Pr} = 6.1989$ and $Ste = 0.1$ agrees well with the results of scaling law correlations at different Rayleigh number Ra.

3.4 GPU acceleration

The characteristic of highly parallel nature is a significant advantage of LBM over other traditional macroscopic numerical methods. In the CUDA programming platform, the CPU and GPU work as the host and the devices, respectively, as shown in **Figure 11** [146]. It means that the parallel tasks are executed on GPU, while the CPU is responsible for the initial conditions and all the sequential

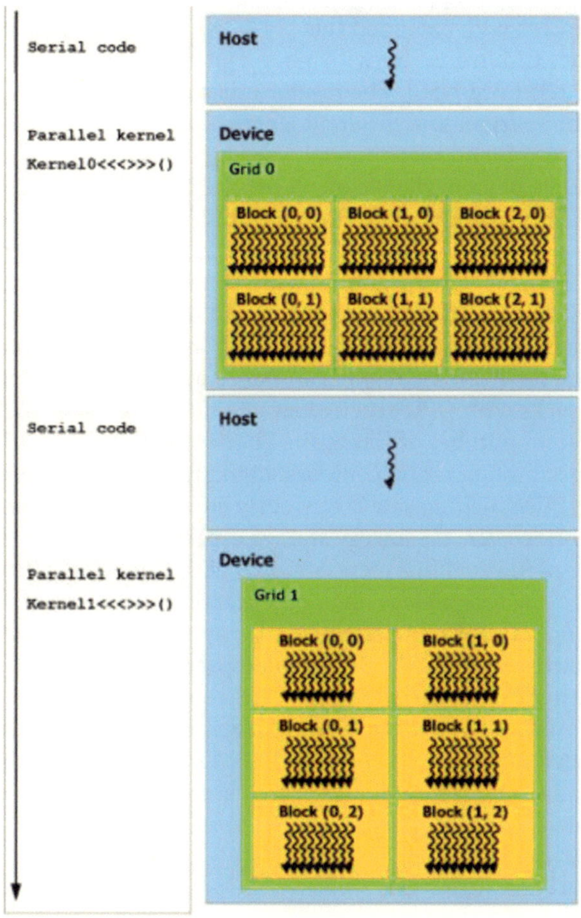

Figure 11.
Schematic diagram of CUDA platform [146].

commands. First, the initial conditions are set up in the host memory, and then the data are moved to the memory of GPU. The threads are grouped into blocks, which are the component of grids as displayed in **Figure 11**. For instance, in the real simulation, if the lattice grid number is (M_x, M_y, M_z) and the block size is (N_x, N_y, N_z), the corresponding grid size is $(M_x/N_x, M_y/N_y, M_z/N_z)$. Furthermore, a kernel is a function, which is executed on the concurrent threads on GPU. The collision step, streaming step, tackling of boundary conditions, and computation of macroscopic variables are completed in different kernels. It is important to note that all the kernels should be synchronized between CPU and GPU. Finally, the data on the GPU should be copied to the host memory of CPU for printing at any specific time when the results are needed. During the recent years, LBM is demonstrated to be appropriate for GPU computing [147–151], and it has been applied to solve several different physical problems [152–155].

4. Applications of LBM modeling in latent heat thermal energy storage

4.1 Enhancement of PCM performance with fins

As presented in the previous sections, the extended fins could be used to increase the heat transfer depth in LHTES system and accelerate the energy storage efficiency. In the recent years, LBM is applied by several researches to study the solid-liquid phase change of PCMs with extended fins [142, 156, 157]. Jourabian et al. studied the melting process of PCM in a cavity with a horizontal fin heated from the sidewall by enthalpy-based D2Q5 LBM [156]. The results indicated that adding a fin enhances the melting rate for all positions and different lengths compared with the LHTES cavity without fin. They also found that although varying the position of the fin from the bottom to the middle has a negligible effect on melting rate, the melting time is increased once the fin is mounted on the top of LHTES cavity. Talati and Taghilou used an implicit LBM to study the PCM solidification in a rectangular finned container [157]. It was found that the optimum aspect ratio of container for solidification equals to 0.5, and changing the fin material from aluminum to copper has no significant influence on the solidification rate. Ren and Chan applied enthalpy-based MRT LBM to investigate the PCM melting performance in an enclosure with internal fins and finned thick walls with GPU acceleration [142]. The transient PCM melting process with different number of fins at the Fourier number $Fo = 0.15$ is shown in **Figure 12** in terms of temperature contours, liquid fraction, and streamlines. It could be found that the PCM melting rate is obviously enhanced by adding more internal fins in the cavity. However, the energy storage capacity of LHTES system is reduced when the number of internal fins increases, so that the appropriate fin configuration and number should be designed for engineering applications. They found that using a less number of longer fins is more effective than applying shorter fins for enhancing the thermal performance of PCMs. Besides, compared with the LHTES cavity with horizontal fins heated from side walls, the LHTES enclosure using vertical fins heated from the bottom surface has a better charging rate. From the above researches, it should be concluded that the enthalpy-based LBM is successful for simulating the conjugate heat transfer with solid-liquid phase change for melting and solidification of PCMs accelerated with fins.

4.2 Nanoparticle-enhanced PCM

The nanoparticles are commonly used to ameliorate the low thermal conductivity of most PCMs. With some appropriate assumptions as presented in Section 2,

Figure 12.
Transient PCM melting process with different number of fins [142].

the thermal performance of PCMs with nanoparticles could be modeled using enthalpy-based LBM. Jourabian et al. studied the convective melting process of Cu-nanoparticle enhanced water-ice in a cylindrical-horizontal annulus [158, 159]. It was demonstrated that the melting rate of NEPCM is accelerated due to the improvement of thermal conductivity and the reduced latent heat. When the heated cylinder located in the bottom section of the annulus, the increment effect on NEPCM melting rate by adding more nanoparticles decreases because of the augmentation of NEPCM viscosity, which weakens the convective heat transfer. Feng et al. investigated the melting of water (ice)-copper nanoparticle NEPCM in a bottom-heated rectangular cavity by treating the latent heat source term with an implicit scheme [122]. They also found that the heat transfer rate of NEPCM increases with respect to the increment of nanoparticle volume fractions. However, the energy storage rate is the most significant parameter for a LHTES unit. Although adding nanoparticles into PCMs could increase their thermal conductivity, it increases the viscosity of PCM, which weakens the convective heat transfer. Due to this reason, the energy storage rate of LHTES system may even decrease with the increasing nanoparticle volume fractions, especially for the case with large temperature gradient. On the other hand, the increment of nanoparticle volume fraction decreases the energy storage capacity of LHTES unit, so that the energy storage rate could also be affected. Under this circumstance, more future research attentions should be paid on the influences of nanoparticles on the energy storage rate of NEPCM. In addition, the solid-liquid phase change model of NEPCM using LBM could be extended to study the charging and discharging of NEPCM under hybrid heat transfer enhancement techniques such as nanoparticle-fin or nanoparticle-metal foam combinations.

4.3 PCM filled with metal foams

Due to its intersected and connected heat transfer channels, metal foam is inserted into the LHTES system for enhancing the melting and solidification rates of PCMs. The numerical modeling of PCM solid-liquid phase change phenomenon filled in metal foams using enthalpy-based LBM could be categorized as the representative elementary volume (REV) scale modeling [128–130, 160] and pore-scale modeling [111, 161]. Tao et al. investigated the performance of metal foams/paraffin composite PCM in a LHTES cavity using LBM at REV scale [160]. They found that increasing the metal foam PPI (number of pores per inch) could enhance the conduction heat transfer, while the convective heat transfer is weakened. In addition, although decreasing the metal foam porosity could accelerate the PCM melting rate, the energy storage density of LHTES unit is dramatically reduced. Due to the above two tradeoffs, the optimum metal foam structure with the porosity of 0.94 and PPI of 45 is highly recommended. However, the REV scale modeling requires the use of some empirical relations, and the influences of metal foam morphology on energy storage rate are difficult to be analyzed. As a comparison, with the advantage of

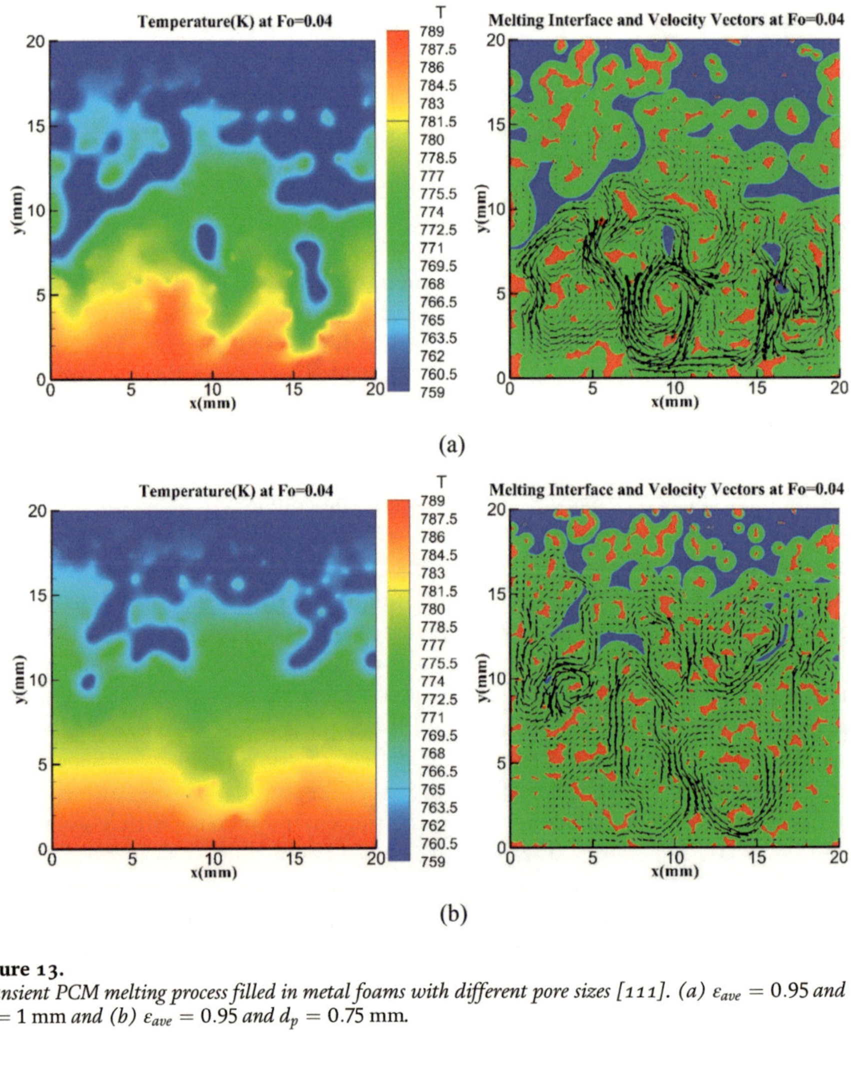

Figure 13.
Transient PCM melting process filled in metal foams with different pore sizes [111]. (a) $\varepsilon_{ave} = 0.95$ *and* $d_p = 1$ mm *and (b)* $\varepsilon_{ave} = 0.95$ *and* $d_p = 0.75$ mm.

LBM for tackling complex boundary conditions, pore-scale modeling without using any empirical equations achieves the detailed fluid flow and heat transfer information inside the metal foams, so that the energy storage efficiency of LHTES unit could be further optimized. Ren et al. investigated the effects of metal foam characteristics on the PCM melting rate through pore-scale LBM modeling [111]. The transient PCM melting process inside the metal foams of different pore sizes at the Fourier number $Fo = 0.04$ is plotted with respect to temperature field, melting interface, and fluid velocity vectors in **Figure 13**. It could be observed that the charging rate of PCM is accelerated when the pore size decreases from $d_p = 1$ mm to $d_p = 0.75$ mm at a porosity of $\varepsilon_{ave} = 0.95$. Furthermore, the temperature field in the melted region of LHTES with smaller pore size $d_p = 0.75$ mm is found to be more uniform than that of LHTES with pore size $d_p = 1.0$ mm, which means that the thermal conductivity and its corresponding conduction heat transfer in LHTES unit are improved with a decreasing pore size (increasing PPI). Besides, they also concluded that an appropriate metal foam porosity should be chosen in the real engineering applications in order to balance the PCM melting speed and the energy storage density of LHTES unit. By reconstructing the microstructure of metal foam with X-ray computed tomography, Li et al. investigated the solid-liquid phase change phenomenon of PCM inserted with metal foams under different gravitational acceleration conditions [161]. The results indicated that the transition of the dominant heat transfer mechanism from convection to conduction occurs when the gravity gradually decreases. Due to the attenuated convection effect with decreasing gravity, the PCM melting development is dramatically hindered. Besides, they also concluded that the decreasing metal foam porosity could enhance the effective thermal conductivity of LHTES unit because of the extended heat transfer area. However, the above pore-scale modeling of PCM charging and discharging processes enhanced by metal foams is limited in two-dimensional cases. The three-dimensional pore-scale modeling, which is indispensable for optimizing the complicated metal foam structures, should be carried out in the future work.

4.4 PCM with heat pipes

The heat pipes are used to transfer heat between PCMs and heat transfer fluid (HTF), so that the charging and discharging of LHTES system could be accelerated. The configuration, arrangement, and number of heat pipes in the LHTES unit are essential for the PCM melting and solidification speed. Luo et al. applied LBM to study the convection melting in complex LHTES system with heat tubes [162]. The effects of inner heat pipe arrangement on PCM melting process are illustrated in **Figure 14** with respect to temperature, flow, and phase fields at the Fourier number $Fo = 3$, the Prandtl number $\Pr = 0.2$, the Stefan number $Ste = 0.02$, and the Rayleigh number $Ra = 5 \times 10^4$. For the case using centrosymmetric inner heat pipes, the conduction heat transfer is dominant because the inner heat pipe is surrounded by solid PCM at the melting temperature, so that its melting rate is faster than the LHTES system with inline or staggered heat tubes as displayed in **Figure 15**. In addition, there is no obvious difference between the melting rates of LHTES systems using inline or staggered heat pipes. Although using heat pipe individually could enhance the thermal performance of LHTES to some extent, other heat transfer enhancement techniques are usually coupled with heat pipes such as fins or nanoparticles to further improve the PCM charging and discharging rate.

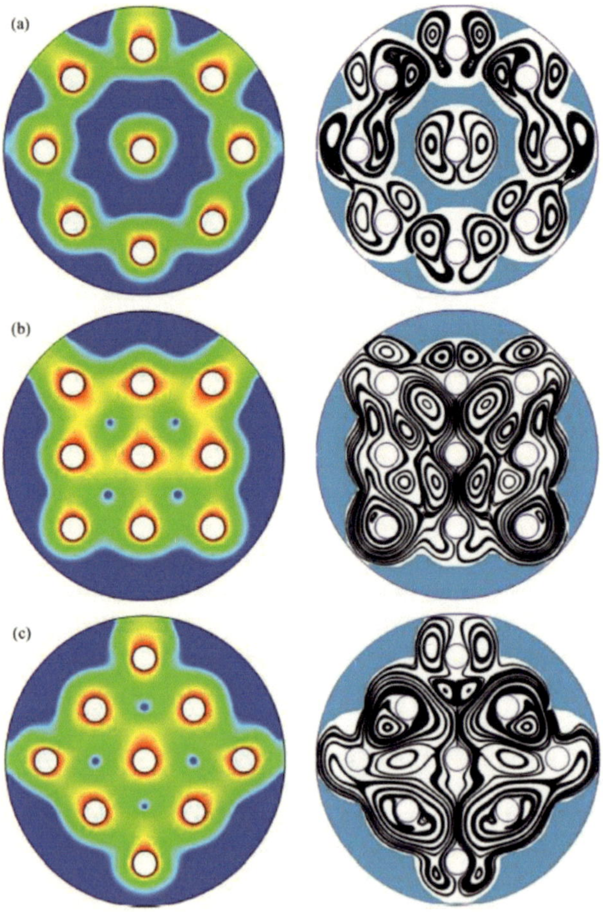

Figure 14.
Temperature, flow, and phase fields at Fo = 3 *with different arrangements of inner tubes: (a) centrosymmetric, (b) inline, and (c) staggered for the Prandtl number* Pr = 0.2*, the Stefan number Ste* = 0.02*, and the Rayleigh number Ra* = 5×10^4 *[162].*

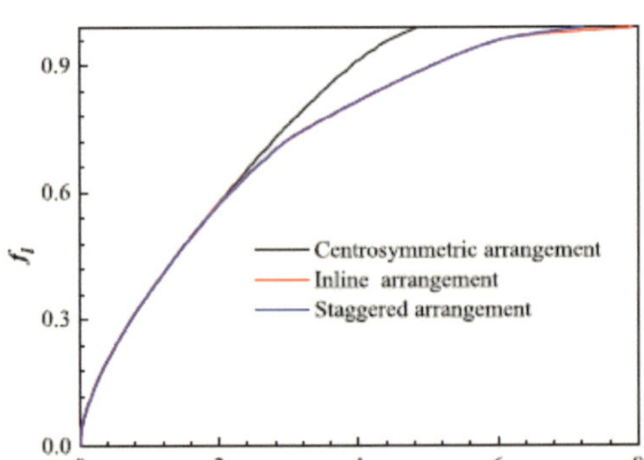

Figure 15.
The total liquid fraction f_l *versus the Fourier number Fo for different arrangements of inner tubes at the Prandtl number* Pr = 0.2*, the Stefan number Ste* = 0.02*, and the Rayleigh number Ra* = 5×10^4 *[162].*

4.5 PCM with hybrid heat transfer enhancement techniques

In this section, the melting and solidification of PCMs enhanced using combined heat transfer enhancement techniques modeled by LBM are discussed, and the effectiveness of different approaches for improving the heat transfer capability of LHTES unit is compared. Huo and Rao carried out an investigation of NEPCM solid liquid phase change process in a cavity with a heat pipe and separate plates using LBM [163]. It was found that the NEPCM of the case with separate located in the middle of cavity melts fastest because of the weakened heat accumulation. The results also showed that when the location of separate is less than 0.3, the melting rate of NEPCM is even slowed down due to the heat accumulation around the separate plate. Gao et al. developed an enthalpy-based MRT LBM model with a free parameter in the equilibrium distribution function for solid-liquid phase change in porous media and conjugate heat transfer with high-computational efficiency and stability [164]. Then, they investigated the PCM melting process in porous media with a conducting fin, and the results indicated that the heat transfer rate could be further improved by adding a fin into the porous medium. It was also observed that the vertical position of the fin has no remarkable impact on the PCM melting speed when there exists porous media. Jourabian et al. investigated the constrained ice melting around a cylinder using three heat transfer enhancement techniques by LBM [165]. They pointed out that adding nanoparticles may increase the dynamic viscosity of the base PCM, which

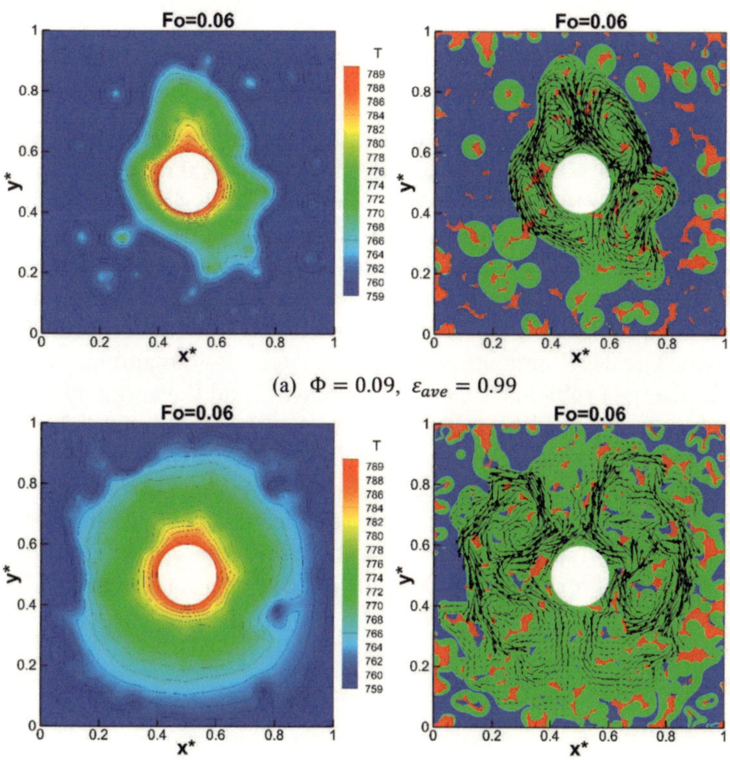

(a) $\Phi = 0.09,\ \varepsilon_{ave} = 0.99$

(b) $\Phi = 0.01,\ \varepsilon_{ave} = 0.91$

Figure 16.
Transient nanoparticle-enhanced PCM melting process filled in metal foams of pore size $d_p = 1$ mm and heat pipe radius $R = 2$ mm [166]. (a) $\Phi = 0.09$, $\varepsilon_{ave} = 0.99$ and (b) $\Phi = 0.01$, $\varepsilon_{ave} = 0.91$.

has a negative effect on natural convective heat transfer. Besides, the tradeoff between consolidating the conduction heat transfer and weakening the convective heat transfer through decreasing the metal foam porosity needs further attention and investigation. Ren et al. completed a pore-scale comparative study of nanoparticle-enhanced PCM melting in a heat pipe–assisted LHTES unit with metal foams [166]. The melting process of NEPCM in a LHTES cavity filled with metal foams using a heat pipe at the Fourier number $Fo = 0.06$, pore size $d_p = 1$ mm, and heat pipe radius $R = 2$ mm is shown in **Figure 16**. Although different combinations of nanoparticle volume fraction Φ and metal foam porosity ε_{ave} are used for the LHTES unit, it should be noted that the volume fraction of pure PCM is kept at 90%, so that the energy storage capacity is unchanged for a fair comparison. It could be observed that the NEPCM with less nanoparticle volume fraction and lower metal foam porosity melts faster than that with more nanoparticles in the higher porosity metal foam. This finding indicated that using metal foams is more effective than adding nanoparticles for enhancing the charging rate of NEPCMs. They also found that there exists an optimum heat pipe radius for achieving the best energy storage rate in LHTES unit. As discussed in the above sections, LBM is demonstrated to be appropriate for simulating the charging and discharging processes of PCMs in LHTES system with different kinds of heat transfer enhancement technologies. Under this circumstance, the enthalpy-based LBM will definitely play a more significant role in the future research of thermal energy storage using PCMs.

5. Summary

In this chapter, different heat transfer enhancement techniques of PCMs for LHTES unit are discussed and compared. As the numerical modeling plays a significant role in clarifying the mechanism of complicated physical processes, the mathematical models for fluid flow of liquid PCMs and the PCM solid-liquid phase change phenomenon are presented. In order to investigate the PCM charging and discharging processes enhanced by nanoparticles, fins, or metal foams, the empirical relations for nanofluids and the mathematical formula for conjugate heat transfer are shown. The development history of the lattice Boltzmann method for solid-liquid phase change problems is carefully reviewed, and the enthalpy-based multiple-relaxation-time LBM is discussed in detail due to its simplicity and robustness. Besides, the implementation of GPU computing is briefly discussed to accelerate the computational efficiency of LBM modeling. Then, the applications of LBM modeling in LHTES system with different heat transfer enhancement approaches are presented, which demonstrate that the mesoscopic and highly parallel LBM is powerful for understanding the melting and solidification processes of PCMs.

Acknowledgements

This work was financially sponsored by the National Key Research and Development Program of China (No. 2017YFB0102703), the National Natural Science Foundation of China (No. 51536003), 111 Project (grant number B16038), and the Foundation for Innovative Research Groups of the National Natural Science Foundation of China (No. 51721004).

Author details

Zhiguo Qu
Key Laboratory of Thermo-Fluid Science Engineering of MOE, School of Energy and Power Engineering, Xi'an Jiaotong University, Xi'an, Shaanxi, China

*Address all correspondence to: zgqu@mail.xjtu.edu.cn

References

[1] Chen H, Cong TN, Yang W, Tan C, Li Y, Ding Y. Progress in electrical energy storage system: A critical review. Progress in Natural Science. 2009;**19**(3): 291-312

[2] Luo X, Wang J, Dooner M, Clarke J. Overview of current development in electrical energy storage technologies and the application potential in power system operation. Applied Energy. 2015; **137**:511-536

[3] Ibrahim NI, Al-Sulaiman FA, Rahman S, Yilbas BS, Sahin AZ. Heat transfer enhancement of phase change materials for thermal energy storage applications: A critical review. Renewable and Sustainable Energy Reviews. 2017;**74**:26-50

[4] Abdulateef AM, Mat S, Abdulateef J, Sopian K, Al-Abidi AA. Geometric and design parameters of fins employed for enhancing thermal energy storage systems: A review. Renewable and Sustainable Energy Reviews. 2018;**82**: 1620-1635

[5] Sharifi N, Bergman TL, Faghri A. Enhancement of PCM melting in enclosures with horizontally-finned internal surfaces. International Journal of Heat and Mass Transfer. 2011;**54**: 4182-4192

[6] Ye W-B, Zhu D-S, Wang N. Numerical simulation on phase-change thermal storage/release in a plate-fin unit. Applied Thermal Engineering. 2011;**31**:3871-3884

[7] Mosaffa AH, Talati F, Rosen MA, Tabrizi HB. Approximate analytical model for PCM solidification in a rectangular finned container with convective cooling boundaries. International Communications in Heat and Mass Transfer. 2012;**39**:318-324

[8] Mat S, Al-Abidi AA, Sopian K, Sulaiman MY, Mohammad AT. Enhance heat transfer for PCM melting in triplex tube with internal-external fins. Energy Conversion and Management. 2013;**74**: 223-236

[9] Al-Abidi AA, Mat S, Sopian K, Sulaiman MY, Mohammad AT. Numerical study of PCM solidification in a triplex tube heat exchanger with internal and external fins. International Journal of Heat and Mass Transfer. 2013;**61**:684-695

[10] Levin PP, Shitzer A, Hetsroni G. Numerical optimization of a PCM-based heat sink with internal fins. International Journal of Heat and Mass Transfer. 2013;**61**:638-645

[11] Fan L-W, Xiao Y-Q, Zeng Y, Fang X, Wang X, Xu X, et al. Effects of melting temperature and the presence of internal fins on the performance of a phase change material (PCM)-based heat sink. International Journal of Thermal Sciences. 2013;**70**:114-126

[12] Kamkari B, Shokouhmand H. Experimental investigation of phase change material melting in rectangular enclosures with horizontal partial fins. International Journal of Heat and Mass Transfer. 2014;**78**:839-851

[13] Sciacovelli A, Gagliardi F, Verda V. Maximization of performance of a PCM latent heat storage system with innovative fins. Applied Energy. 2015; **137**:707-715

[14] Rathod MK, Banerjee J. Thermal performance enhancement of shell and tube latent heat storage unit using longitudinal fins. Applied Thermal Engineering. 2015;**75**:1084-1092

[15] Pakrouh R, Hosseini MJ, Ranjbar AA, Bahrampoury R. A numerical method for PCM-based pin fin heat sinks optimization. Energy Conversion and Management. 2015;**103**:542-552

[16] Jmal I, Baccar M. Numerical study of PCM solidification in a finned tube thermal storage including natural convection. Applied Thermal Engineering. 2015;**84**:320-330

[17] Kalbasi R, Salimpour MR. Constructal design of horizontal fins to improve the performance of phase change material rectangular enclosures. Applied Thermal Engineering. 2015;**91**: 234-244

[18] Yuan Y, Cao X, Xiang B, Du Y. Effect of installation angle of fins on melting characteristics of annular unit for latent heat thermal energy storage. Solar Energy. 2016;**136**:365-378

[19] Eslamnezhad H, Rahimi AB. Enhance heat transfer for phase-change materials in triplex tube heat exchanger with selected arrangements of fins. Applied Thermal Engineering. 2017;**113**: 813-821

[20] Mahdi JM, Lohrasbi S, Ganji DD, Nsofor EC. Accelerated melting of PCM in energy storage systems via novel configuration of fins in the triplex-tube heat exchanger. International Journal of Heat and Mass Transfer. 2018;**124**: 663-676

[21] Ji C, Qin Z, Low Z, Dubey S, Choo FH, Duan F. Non-uniform heat transfer suppression to enhance PCM melting by angled fins. Applied Thermal Engineering. 2018;**129**:269-279

[22] Zhao CY. Review on thermal transport in high porosity cellular metal foams with open cells. International Journal of Heat and Mass Transfer. 2012;**55**:3618-3632

[23] Xiao X, Zhang P, Li M. Preparation and thermal characterization of paraffin/metal foam composite phase change material. Applied Energy. 2013; **112**:1357-1366

[24] Feng S, Zhang Y, Shi M, Wen T, Lu TJ. Unidirectional freezing of phase change materials saturated in open-cell metal foams. Applied Thermal Engineering. 2015;**88**:315-321

[25] Zhao CY, Wu ZG. Heat transfer enhancement of high temperature thermal energy storage using metal foams and expanded graphite. Solar Energy Materials & Solar Cells. 2011;**95**: 636-643

[26] Tian Y, Zhao CY. A numerical investigation of heat transfer in phase change materials (PCMs) embedded in porous metals. Energy. 2011;**36**: 5539-5546

[27] Li WQ, Qu ZG, He YL, Tao WQ. Experimental and numerical studies on melting phase change heat transfer in open-cell metallic foams with paraffin. Applied Thermal Engineering. 2012;**37**: 1-9

[28] Zhou D, Zhao CY. Experimental investigations on heat transfer in phase change materials (PCMs) embedded in porous materials. Applied Thermal Engineering. 2011;**31**:970-977

[29] Cui HT. Experimental investigation on the heat charging process by paraffin filled with high porosity copper foam. Applied Thermal Engineering. 2012;**39**: 26-28

[30] Qu ZG, Li WQ, Wang JL, Tao WQ. Passive thermal management using metal foam saturated with phase change material in a heat sink. International Communications in Heat and Mass Transfer. 2012;**39**:1546-1549

[31] Tian Y, Zhao CY. Thermal and exergetic analysis of metal foam-enhanced cascaded thermal energy storage (MF-CTES). International Journal of Heat and Mass Transfer. 2013;**58**:86-96

[32] Baby R, Balaji C. Experimental investigations on thermal performance enhancement and effect of orientation on porous matrix filled PCM based heat

sink. International Journal of Communications in Heat and Mass Transfer. 2013;**46**:27-30

[33] Liu Z, Yao Y, Wu H. Numerical modelling for solid-liquid phase change phenomena in porous media: Shell-and-tube type latent heat thermal energy storage. Applied Energy. 2013;**112**: 1222-1232

[34] Chen Z, Gao D, Shi J. Experimental and numerical study on melting of phase change materials in metal foams at pore scale. International Journal of Heat and Mass Transfer. 2014;**72**: 646-655

[35] Sundarram SS, Li W. The effect of pore size and porosity on thermal management performance of phase change material infiltrated microcellular metal foams. Applied Thermal Engineering. 2014;**64**:147-154

[36] Xiao X, Zhang P, Li M. Effective thermal conductivity of open-cell metal foams impregnated with pure paraffin for latent heat storage. International Journal of Thermal Sciences. 2014;**81**: 94-105

[37] Mancin S, Diani A, Doretti L, Hooman K, Rossetto L. Experimental analysis of phase change phenomenon of paraffin waxes embedded in copper foams. International Journal of Thermal Sciences. 2015;**90**:79-89

[38] Deng Z, Liu X, Zhang C, Huang Y, Chen Y. Melting behaviors of PCM in porous metal foam characterized by fractal geometry. International Journal of Heat and Mass Transfer. 2017;**113**: 1031-1042

[39] Zhang P, Meng ZN, Zhu H, Wang YL, Peng SP. Melting heat transfer characteristics of a composite phase change material fabricated by paraffin and metal foam. Applied Energy. 2017; **185**:1971-1983

[40] Zhang Z, Cheng J, He X. Numerical simulation of flow and heat transfer in composite PCM on the basis of two different models of open-cell metal foam skeletons. International Journal of Heat and Mass Transfer. 2017;**112**: 959-971

[41] Yang X, Feng S, Zhang Q, Chai Y, Jin L, Lu TJ. The role of porous metal foam on the unidirectional solidification of saturating fluid for cold storage. Applied Energy. 2017;**194**:508-521

[42] Esapour M, Hamzehnezhad A, Rabienataj Darzi AA, Jourabian M. Melting and solidification of PCM embedded in porous metal foam in horizontal multi-tube heat storage system. Energy Conversion and Management. 2018;**171**:398-410

[43] Zhu Z-Q, Huang Y-K, Hu N, Zeng Y, Fan L-W. Transient performance of a PCM-based heat sink with a partially filled metal foam: Effects of the filling height ratio. Applied Thermal Engineering. 2018;**128**:966-972

[44] Zheng H, Wang C, Liu Q, Tian Z, Fan X. Thermal performance of copper foam/paraffin composite phase change material. Energy Conversion and Management. 2018;**157**:372-381

[45] Ji H, Sellan DP, Pettes MT, Kong X, Ji J, Shi L, et al. Enhanced thermal conductivity of phase change materials with ultrathin-graphite foams for thermal energy storage. Energy & Environmental Science. 2014;7: 1185-1192

[46] Li M. A nano-graphite/paraffin phase change material with high thermal conductivity. Applied Energy. 2013;**106**:25-30

[47] Alshaer WG, Nada SA, Rady MA, Le Bot C, Palomo Del Barrio E. Numerical investigation of using carbon foam/ PCM/nano carbon tubes composites in thermal management of electronic

equipment. Energy Conversion and Management. 2015;**89**:873-884

[48] Zou D, Ma X, Liu X, Zheng P, Hu Y. Thermal performance enhancement of composite phase change materials (PCM) using graphene and carbon nanotubes as additives for the potential application in lithium-ion power battery. International Journal of Heat and Mass Transfer. 2018;**120**:33-41

[49] Khodadadi JM, Fan L, Babaei H. Thermal conductivity enhancement of nanostructure-based colloidal suspensions utilized as phase change materials for thermal energy storage: A review. Renewable and Sustainable Energy Reviews. 2013;**24**:418-444

[50] Khodadadi JM, Hosseinizadeh SF. Nanoparticle-enhanced phase change materials (NEPCM) with great potential for improved thermal energy storage. International Communications in Heat and Mass Transfer. 2007;**34**:534-543

[51] Chieruzzi M, Cerritelli GF, Miliozzi A, Kenny JM. Effect of nanoparticles on heat capacity of nanofluids based on molten salts as PCM for thermal energy storage. Nanoscale Research Letters. 2013;**8**:448

[52] Kibria MA, Anisur MR, Mahfuz MH, Saidur R, Metselaar IHSC. A review on thermophysical properties of nanoparticle dispersed phase change materials. Energy Conversion and Management. 2015;**95**:69-89

[53] Arasu AV, Mujumdar AS. Numerical study on melting of paraffin wax with Al_2O_3 in a square enclosure. International Communications in Heat and Mass Transfer. 2012;**39**:8-16

[54] Fan L, Khodadadi JM. An experimental investigation of enhanced thermal conductivity and expedited unidirectional freezing of cyclohexane-based nanoparticle suspensions utilized as nano-enhanced phase change materials (NePCM). International Journal of Thermal Sciences. 2012;**62**: 120-126

[55] Sciacovelli A, Colella F, Verda V. Melting of PCM in a thermal energy storage unit: Numerical investigation and effect of nanoparticle enhancement. International Journal of Energy Research. 2013;**37**:1610-1623

[56] Dhaidan NS, Khodadadi JM, Al-Hattab TA, Al-Mashat SM. Experimental and numerical investigation of melting of phase change material/nanoparticle suspensions in a square container subjected to a constant heat flux. International Journal of Heat and Mass Transfer. 2013;**66**:672-683

[57] Dhaidan NS, Khodadadi JM, Al-Hattab TA, Al-Mashat SM. Experimental and numerical study of constrained melting of n-octadecane with CuO nanoparticle dispersions in a horizontal cylindrical capsule subjected to a constant heat flux. International Journal of Heat and Mass Transfer. 2013;**67**:523-534

[58] Zeng Y, Fan L-W, Xiao Y-Q, Yu Z-T, Cen K-F. An experimental investigation of melting of nanoparticle-enhanced phase change materials (NePCMs) in a bottom-heated vertical cylindrical cavity. International Journal of Heat and Mass Transfer. 2013;**66**: 111-117

[59] Chandrasekaran P, Cheralathan M, Kumaresan V, Velraj R. Enhanced heat transfer characteristics of water based copper oxide nanofluid PCM (phase change material) in a spherical capsule during solidification for energy efficient cool thermal storage system. Energy. 2014;**72**:636-642

[60] Altohamy AA, Abd Rabbo MF, Sakr RY, Attia AAA. Effect of water based Al_2O_3 nanoparticle PCM on cool storage performance. Applied Thermal Engineering. 2015;**84**:331-338

[61] Sathishkumar A, Kumaresan V, Velraj R. Solidification characteristics of water based graphene nanofluid PCM in a spherical capsule for cool thermal energy storage applications. International Journal of Refrigeration. 2016;**66**:73-83

[62] Arici M, Tutuncu E, Kan M, Karabay H. Melting of nanoparticle-enhanced paraffin wax in a rectangular enclosure with partially active walls. International Journal of Heat and Mass Transfer. 2017;**104**:7-17

[63] Parsazadeh M, Duan X. Numerical and statistical study on melting of nanoparticle enhanced phase change material in a shell-and-tube thermal energy storage system. Applied Thermal Engineering. 2017;**111**: 950-960

[64] Pahamli Y, Hosseini MJ, Ranjbar AA, Bahrampoury R. Effect of nanoparticle dispersion and inclination angle on melting of PCM in a shell and tube heat exchanger. Journal of the Taiwan Institute of Chemical Engineers. 2017;**81**:316-334

[65] Motahar S, Alemrajabi AA, Khodabandeh R. Experimental study on solidification process of a phase change material containing TiO_2 nanoparticles for thermal energy storage. Energy Conversion and Management. 2017;**138**: 162-170

[66] Elbahjaoui R, El Qarnia H. Transient behavior analysis of the melting of nanoparticle-enhanced phase change material inside a rectangular latent heat storage unit. Applied Thermal Engineering. 2017;**112**:720-738

[67] Wang F, Zhang C, Liu J, Fang X, Zhang Z. Highly stable graphite nanoparticle-dispersed phase change emulsions with little supercooling and high thermal conductivity for cold energy storage. Applied Energy. 2017; **188**:97-106

[68] Sheikholeslami M, Ghasemi A, Li Z, Shafee A, Saleem S. Influence of CuO nanoparticles on heat transfer behavior of PCM in solidification process considering radiative source term. International Journal of Heat and Mass Transfer. 2018;**126**:1252-1264

[69] Sheikholeslami M. Finite element method for PCM solidification in existence of CuO nanoparticles. Journal of Molecular Liquids. 2018;**265**:347-355

[70] Sheikholeslami M. Solidification of NEPCM under the effect of magnetic field in a porous thermal energy storage enclosure using CuO nanoparticles. Journal of Molecular Liquids. 2018;**263**: 303-315

[71] Sheikholeslami M, Ghasemi A. Solidification heat transfer of nanofluid in existence of thermal radiation by means of FEM. International Journal of Heat and Mass Transfer. 2018;**123**: 418-431

[72] Li Z, Sheikholeslami M, Samandari M, Shafee A. Nanofluid unsteady heat transfer in a porous energy storage enclosure in existence of Lorentz forces. International Journal of Heat and Mass Transfer. 2018;**127**:914-926

[73] Sheikholeslami M, Li Z, Shafee A. Lorentz force effect on NEPCM heat transfer during solidification in a porous energy storage system. International Journal of Heat and Mass Transfer. 2018;**127**:665-674

[74] Naghavi MS, Ong KS, Mehrali M, Badruddin IA, Metselaar HSC. A state-of-the-art review on hybrid heat pipe latent heat storage systems. Energy Conversion and Management. 2015;**105**: 1178-1204

[75] Tiari S, Mahdavi M, Qiu S. Experimental study of a latent heat thermal energy storage system assisted by a heat pipe network. Energy

Conversion and Management. 2017;**153**: 362-373

[76] Robak CW, Bergman TL, Faghri A. Enhancement of latent heat energy storage using embedded heat pipes. International Journal of Heat and Mass Transfer. 2011;**54**:3476-3484

[77] Liu X, Fang G, Chen Z. Dynamic charging characteristics modeling of heat storage device with heat pipe. Applied Thermal Engineering. 2011;**31**: 2902-2908

[78] Xiaohong G, Bin L, Yongxian G, Xiugan Y. Two-dimensional transient study thermal analysis of PCM canister of a heat pipe receiver under microgravity. Applied Thermal Engineering. 2011;**31**:735-741

[79] Nithyanandam K, Pitchumani R. Analysis and optimization of a latent thermal energy storage system with embedded heat pipes. International Journal of Heat and Mass Transfer. 2011; **54**:4596-4610

[80] Sharifi N, Wang S, Bergman TL, Faghri A. Heat pipe-assisted melting of a phase change material. International Journal of Heat and Mass Transfer. 2012;**55**:3458-3469

[81] Shabgard H, Robak CW, Bergman TL, Faghri A. Heat transfer and exergy analysis of cascaded latent heat storage with gravity-assisted heat pipes for concentrating solar power applications. Solar Energy. 2012;**86**:816-830

[82] Nithyanandam K, Pitchumani R. Computational studies on a latent thermal energy storage system with integral heat pipes for concentrating solar power. Applied Energy. 2013;**103**: 400-415

[83] Sharifi N, Bergman TL, Allen MJ, Faghri A. Melting and solidification enhancement using a combined heat pipe, foil approach. International Journal of Heat and Mass Transfer. 2014;**78**:930-941

[84] Jung EG, Boo JH. Thermal analytical model of latent thermal storage with heat pipe heat exchanger for concentrated solar power. Solar Energy. 2014;**102**:318-332

[85] Lorente S, Bejan A, Niu JL. Phase change heat storage in an enclosure with vertical pipe in the center. International Journal of Heat and Mass Transfer. 2014;**72**:329-335

[86] Tiari S, Qiu S, Mahdavi M. Numerical study of finned heat pipe-assisted thermal energy storage system with high temperature phase change material. Energy Conversion and Management. 2015;**89**:833-842

[87] Sharifi N, Faghri A, Bergman TL, Andraka CE. Simulation of heat pipe-assisted latent heat thermal energy storage with simultaneous charging and discharging. International Journal of Heat and Mass Transfer. 2015;**80**:170-179

[88] Liu Z-h, Zheng B-c, Wang Q, Li S-S. Study on the thermal storage performance of a gravity-assisted heat-pipe thermal storage unit with granular high-temperature phase-change materials. Energy. 2015;**81**:754-765

[89] Zhao J, Rao Z, Liu C, Li Y. Experimental investigation on thermal performance of phase change material coupled with closed-loop oscillating heat pipe (PCM/CLOHP) used in thermal management. Applied Thermal Engineering. 2016;**93**:90-100

[90] Tiari S, Qiu S, Mahdavi M. Discharging process of a finned heat pipe-assisted thermal energy storage system with high temperature phase change material. Energy Conversion and Management. 2016;**118**:426-437

[91] Zhao J, Rao Z, Liu C, Li Y. Experiment study of oscillating heat

pipe and phase change materials coupled for thermal energy storage and thermal management. International Journal of Heat and Mass Transfer. 2016;**99**:252-260

[92] Motahar S, Khodabandeh R. Experimental study on the melting and solidification of a phase change material enhanced by heat pipe. International Communications in Heat and Mass Transfer. 2016;**73**:1-6

[93] Amini A, Miller J, Jouhara H. An investigation into the use of the heat pipe technology in thermal energy storage heat exchangers. Energy. 2017; **136**:163-172

[94] Naghavi MS, Ong KS, Badruddin IA, Mehrali M, Metselaar HSC. Thermal performance of a compact design heat pipe solar collector with latent heat storage in charging/discharging modes. Energy. 2017;**127**:101-115

[95] Pan C, Vermaak N, Romero C, Neti S, Hoenig S, Chen C-H. Efficient optimization of a longitudinal finned heat pipe structure for a latent thermal energy storage system. Energy Conversion and Management. 2017;**153**: 93-105

[96] Zhao J, Qu J, Rao Z. Thermal characteristic and analysis of closed loop oscillation heat pipe/phase change material (CLOHP/PCM) coupling module with different working media. International Journal of Heat and Mass Transfer. 2018;**126**:257-266

[97] Sheikholeslami M. Numerical modeling of nano enhanced PCM solidification in an enclosure with metallic fin. Journal of Molecular Liquids. 2018;**259**:424-438

[98] Rabienataj Darzi AA, Jourabian M, Farhadi M. Melting and solidification of PCM enhanced by radial conductive fins and nanoparticles in cylindrical annulus. Energy Conversion and Management. 2016;**118**:253-263

[99] Lohrasbi S, Sheikholeslami M, Ganji DD. Multi-objective RSM optimization of fin assisted latent heat thermal energy storage system based on solidification process of phase change material in presence of copper nanoparticles. Applied Thermal Engineering. 2017;**118**: 430-447

[100] Parsazadeh M, Duan X. Numerical study on the effects of fins and nanoparticles in a shell and tube phase change thermal energy storage unit. Applied Energy. 2018;**216**:142-156

[101] Hossain R, Mahmud S, Dutta A, Pop I. Energy storage system based on nanoparticle-enhanced phase change material inside porous medium. International Journal of Thermal Sciences. 2015;**91**:49-58

[102] Tasnim SH, Hossain R, Mahmud S, Dutta A. Convection effect on the melting process of nano-PCM inside porous enclosure. International Journal of Heat and Mass Transfer. 2015;**85**: 206-220

[103] Srivatsa PVSS, Baby R, Balaji C. Numerical investigation of PCM based heat sinks with embedded metal foam/ crossed plate fins. Numerical Heat Transfer, Part A: Applications. 2014;**66**: 1131-1153

[104] Mahdi JM, Nsofor EC. Melting enhancement in triplex-tube latent thermal energy storage system using nanoparticles-fins combination. International Journal of Heat and Mass Transfer. 2017;**109**:417-427

[105] Mahdi JM, Nsofor EC. Solidification enhancement of PCM in a triplex-tube thermal energy storage system with nanoparticles and fins. Applied Energy. 2018;**211**:975-986

[106] Mahdi JM, Nsofor EC. Melting enhancement in triplex-tube latent heat energy storage system using nanoparticles-metal foam combination. Applied Energy. 2017;**191**:22-34

[107] Mahdi JM, Nsofor EC. Solidification enhancement in a triplex-tube latent heat thermal energy storage system using nanoparticles-metal foam combination. Energy. 2017;**126**:501-512

[108] Brinkman HC. The viscosity of concentrated suspensions and solutions. The Journal of Chemical Physics. 1952; **20**(4):571

[109] Kakac S, Pramuanjaroenkij A. Review of convective heat transfer enhancement with nanofluids. International Journal of Heat and Mass Transfer. 2009;**52**:3187-3196

[110] Wang M, Wang J, Pan N, Chen S. Mesoscopic predictions of the effective thermal conductivity for microscale random porous media. Physical Review E. 2007;**75**:036702

[111] Ren Q, He Y-L, Su K-Z, Chan CL. Investigation of the effect of metal foam characteristics on the PCM melting performance in a latent heat thermal energy storage unit by pore-scale lattice Boltzmann modeling. Numerical Heat Transfer, Part A: Applications. 2017;**72**: 745-764

[112] Chen S, Doolen GD. Lattice Boltzmann method for fluid flows. Annual Review of Fluid Mechanics. 1998;**30**:329-364

[113] Chen L, Kang Q, Mu Y, He Y-L, Tao W-Q. A critical review of the pseudopotential multiphase lattice Boltzmann model: Methods and applications. International Journal of Heat and Mass Transfer. 2014;**76**: 210-236

[114] Li Q, Luo KH, Kang QJ, He YL, Chen Q, Liu Q. Lattice Boltzmann methods for multiphase flow and phase-change heat transfer. Progress in Energy and Combustion Science. 2016;**52**: 62-105

[115] Zhang J. Lattice Boltzmann method for microfluidics: Models and applications. Microfluidics and Nanofluidics. 2010;**10**:1-28

[116] Miller W. The lattice Boltzmann method: A new tool for numerical simulation of the interaction of growth kinetics and melt flow. Journal of Crystal Growth. 2001;**230**(1–2):263-269

[117] Rasin I, Miller W, Succi S. Phase-field lattice kinetic scheme for the numerical simulation of dendritic growth. Physical Review E. 2005;**72**: 066705

[118] Huang RZ, Wu HY. An immersed boundary-thermal lattice Boltzmann method for solid-liquid phase change. Journal of Computational Physics. 2014; **277**:305-319

[119] Jiaung WS, Ho JR, Kuo CP. Lattice Boltzmann method for the heat conduction problem with phase change. Numerical Heat Transfer, Part B: Fundamentals. 2001;**39**:167-187

[120] Huber C, Parmigiani A, Chopard B, Manga M, Bachmann O. Lattice Boltzmann model for melting with natural convection. International Journal of Heat and Fluid Flow. 2008; **29**(5):1469-1480

[121] Eshraghi M, Felicelli SD. An implicit lattice Boltzmann model for heat conduction with phase change. International Journal of Heat and Mass Transfer. 2012;**55**(9–10):2420-2428

[122] Feng Y, Li H, Li L, Lin B, Wang T. Numerical investigation on the melting of nanoparticle-enhanced phase change (NEPCM) in a bottom-heated rectangular cavity using lattice Boltzmann method. International

Journal of Heat and Mass Transfer. 2015; **81**:415-425

[123] Huang RZ, Wu HY, Cheng P. A new lattice Boltzmann model for solid-liquid phase change. International Journal of Heat and Mass Transfer. 2013;**59**:295-301

[124] Huang RZ, Wu HY. Phase interface effects in the total enthalpy-based lattice Boltzmann model for solid-liquid phase change. Journal of Computational Physics. 2015;**294**:346-362

[125] Li D, Ren Q, Tong Z-X, He Y-L. Lattice Boltzmann models for axisymmetric solid-liquid phase change. International Journal of Heat and Mass Transfer. 2017;**112**:795-804

[126] Li D, Tong Z-X, Ren Q, He Y-L, Tao W-Q. Three-dimensional lattice Boltzmann models for solid-liquid phase change. International Journal of Heat and Mass Transfer. 2017;**115**:1334-1347

[127] Su Y, Davidson JH. A new mesoscopic scale timestep adjustable non-dimensional lattice Boltzmann method for melting and solidification heat transfer. International Journal of Heat and Mass Transfer. 2016;**92**: 1106-1119

[128] Wu W, Zhang S, Wang S. A novel lattice Boltzmann model for the solid-liquid phase change with the convection heat transfer in the porous media. International Journal of Heat and Mass Transfer. 2017;**104**:675-687

[129] Gao D, Tian F-B, Chen Z, Zhang D. An improved lattice Boltzmann method for solid-liquid phase change in porous media under local thermal non-equilibrium conditions. International Journal of Heat and Mass Transfer. 2017; **110**:58-62

[130] Liu Q, He Y-L, Li Q. Enthalpy-based multiple-relaxation-time lattice Boltzmann method for solid-liquid phase-change heat transfer in metal foams. Physical Review E. 2017;**96**:023303

[131] Yu Z, Fan LS. Multirelaxation-time interaction-potential-based lattice Boltzmann model for two phase flow. Physical Review E. 2010;**82**:046708

[132] Li Q, Luo KH, Li XJ. Lattice Boltzmann modeling of multiphase flows at large density ratio with an improved pseudopotential model. Physical Review E. 2013;**87**:053301

[133] Hu A, Li L, Uddin R. Force method in a pseudo-potential lattice Boltzmann model. Journal of Computational Physics. 2015;**294**:78-89

[134] Huang R, Wu H. Third-order analysis of pseudopotential lattice Boltzmann model for multiphase flow. Journal of Computational Physics. 2016; **327**:121-139

[135] McCracken ME, Abraham J. Multiple-relaxation-time lattice-Boltzmann model for multiphase flow. Physical Review E. 2005;**71**:036701

[136] Guo Z, Zheng C. Analysis of lattice Boltzmann equation for microscale gas flows: Relaxation times, boundary conditions and the Knudsen layer. International Journal of Fluid Dynamics. 2008;**22**:465-473

[137] Lallemand P, Luo LS. Theory of the lattice Boltzmann method: Dispersion, dissipation, isotropy, Galilean invariance, and stability. Physical Review E. 2000;**61**:6546-6562

[138] Huang R, Wu H. Total enthalpy-based lattice Boltzmann method with adaptive mesh refinement for solid-liquid phase change. Journal of Computational Physics. 2016;**315**:65-83

[139] Zhao-Li G, Chu-Guang Z, Bao-Chang S. Non-equilibrium extrapolation method for velocity and pressure boundary conditions in the lattice

Boltzmann method. Chinese Physics. 2002;**11**(4):366-374

[140] Guo Z, Zheng C. An extrapolation method for boundary conditions in lattice Boltzmann method. Physics of Fluids. 2002;**14**(6):2007-2010

[141] Yoshida H, Kobayashi T, Hayashi H, Kinjo T, Washizu H, Fukuzawa K. Boundary condition at a two-phase interface in the lattice Boltzmann method for the convection-diffusion equation. Physical Review E. 2014;**90**:013303

[142] Ren Q, Chan CL. GPU accelerated numerical study of PCM melting process in an enclosure with internal fins using lattice Boltzmann method. International Journal of Heat and Mass Transfer. 2016;**100**:522-535

[143] Hu H, Argyropoulos SA. Mathematical modeling of solidification and melting: A review. Modelling and Simulation in Materials Science and Engineering. 1996;**4**:371-396

[144] Mencinger J. Numerical simulation of melting in two-dimensional cavity using adaptive grid. Journal of Computational Physics. 2004;**198**(1): 243-264

[145] Jany P, Bejan A. Scaling theory of melting with natural convection in an enclosure. International Journal of Heat and Mass Transfer. 1988;**31**:1221-1235

[146] CUDA Toolkit Documentation V9.2, NIVIDIA CORPORATION: https://docs.nvidia.com/cuda/

[147] Zhao Y. Lattice Boltzmann based PDE solver on the GPU. The Visual Computer. 2008;**24**:323-333

[148] Riegel E, Indinger T, Adams NA. Implementation of a Lattice-Boltzmann method for numerical fluid mechanics using the NVIDIA CUDA technology. Computer Science—Research and Development. 2009;**23**:241-247

[149] Obrecht C, Kuznik F, Tourancheau B, Roux J-J. A new approach to the lattice Boltzmann method for graphics processing units. Computers & Mathematcs with Applications. 2011;**61**: 3628-3638

[150] Obrecht C, Kuznik F, Tourancheau B, Roux J-J. Multi-GPU implementation of the lattice Boltzmann method. Computers & Mathematcs with Applications. 2013;**65**:252-261

[151] Habich J, Feichtinger C, Kostler H, Hager G, Wellein G. Performance engineering for the lattice Boltmznan method on GPGPUS: Architectural requirements and performance results. Computers & Fluids. 2013;**80**:276-282

[152] Ren Q, Chan CL. Natural convection with an array of solid obstacles in an enclosure by lattice Boltzmann method on a CUDA computation platform. International Journal of Heat and Mass Transfer. 2016;**93**:273-285

[153] Ren Q, Chan CL. Numerical study of double-diffusive convection in a vertical cavity with Soret and Dufour effects by lattice Boltzmann method on GPU. International Journal of Heat and Mass Transfer. 2016;**93**:538-553

[154] Ren Q, Chan CL. Numerical simulation of a 2D electrothermal pump by lattice Boltzmann method on GPU. Numerical Heat Transfer, Part A: Applications. 2016;**69**:677-693

[155] Ren Q. Investigation of pumping mechanism for non-Newtonian blood flow with AC electrothermal forces in a microchannel by hybrid boundary element method and immersed boundary-lattice Boltzmann method. Electrophoresis. 2018;**39**:1329-1338

[156] Jourabian M, Farhadi M, Sedighi K, Darzi AAR, Vazifeshenas Y. International Journal for Numerical Methods in Fluids. 2012;**70**:313-325

[157] Talati F, Taghilou M. Lattice Boltzmann application on the PCM solidification within a rectangular finned container. Applied Thermal Engineering. 2015;**83**:108-120

[158] Jourabian M, Farhadi M, Rabienataj Darzi AA. Outward melting of ice enhanced by Cu nanoparticles inside cylindrical horizontal annulus: Lattice Boltzmann approach. Applied Mathematical Modelling. 2013;**37**: 8813-8825

[159] Jourabian M, Farhadi M, Sedighi K. On the expedited melting of phase change material (PCM) through dispersion of nanoparticles in the thermal storage unit. Computers and Mathematics with Applications. 2014; **67**:1358-1372

[160] Tao YB, You Y, He YL. Lattice Boltzmann simulation on phase change heat transfer in metal foams/paraffin composite phase change material. Applied Thermal Engineering. 2016;**93**: 476-485

[161] Li X, Ma T, Liu J, Zhang H, Wang Q. Pore-scale investigation of gravity effects on phase change heat transfer characteristics using lattice Boltzmann method. Applied Energy. 2018;**222**: 92-103

[162] Luo K, Yao F-J, Yi H-L, Tan H-P. Lattice Boltzmann simulation of convection melting in complex heat storage systems filled with phase change materials. Applied Thermal Engineering. 2015;**86**:238-250

[163] Huo Y, Rao Z. Lattice Boltzmann investigation on phase change of nanoparticle-enhanced phase change material in a cavity with separate plate. Energy Conversion and Management. 2017;**154**:420-429

[164] Gao D, Chen Z, Zhang D, Chen L. Lattice Boltzmann modeling of melting of phase change materials in porous media with conducting fins. Applied Thermal Engineering. 2017;**118**:315-327

[165] Jourabian M, Farhadi M, Rabienataj Darzi AA. Constrained ice melting around one cylinder in horizontal cavity accelerated using three heat transfer enhancement techniques. International Journal of Thermal Sciences. 2018;**125**: 231-247

[166] Ren Q, Meng F, Guo P. A comparative study of PCM melting process in a heat pipe-assisted LHTES unit enhanced with nanoparticles and metal foams by immersed boundary-lattice Boltzmann method at pore-scale. International Journal of Heat and Mass Transfer. 2018;**121**:1214-1228

Chapter 4

The Laboratorial Research of Two-Phase Free Convection Devices for Cooling of Materials and Industrial Machines

Yakov B. Gorelik and Artur H. Khabitov

Abstract

Refrigeration systems based on free convection (two-phase thermosyphons) are used for cooling equipment units in chemical, nuclear power, and steel-making industries, as well as for thermal stabilization of natural materials with temperature-dependent properties, such as permafrost. Results of laboratory testing are reported for two types of thermosyphons applied mainly to the thermal stabilization of frozen ground: (a) vertical tubes with finning and (b) systems with horizontal evaporation tubes (HET systems). Their uses are currently restricted to relatively small thermal loads, but the effect of the loads on the cooling performance remains poorly investigated. Theoretical analysis of internal and external heat transfer in a vertical thermosyphon provides constraints on the boundary conditions at the evaporator wall, to be used in formulating and solving problems on the temperature regime of frozen ground stabilized with thermosyphons. Comparison of measured and calculated parameters that characterize the operation of a model HET system allows improving the calculation quality by applying the respective corrections.

Keywords: thermosyphons, refrigeration systems, free convection, two-phase flow, hydraulic resistance, cooling effect, experimental modeling, theoretical modeling, permafrost

1. Introduction

Refrigeration systems based on free convection (two-phase thermosyphons) are used for cooling equipment units in chemical, nuclear power, and steel-making industries, as well as for the thermal stabilization of natural materials with temperature-dependent properties, such as permafrost. Thermosyphons (also called "heat pipes" [1]) have different design features and sizes depending on application and operation conditions [1, 2]. Generally, two-phase thermosyphons consist of an evaporator and a condenser. The evaporator has a tightly closed case partly filled with liquid working fluid (coolant) and is placed in immediate vicinity of the cooled equipment or inside the cooled ground. The condenser contacts with a colder ambience (e.g., atmospheric air) and gives up heat. Heat flux to the evaporator boils up the fluid, and the released vapor moves to the condenser driven by the saturation vapor pressure gradient. The vapor condenses in the condenser, releasing heat into the environment, while the liquid phase of the fluid returns into the evaporator driven by gravity, capillary, or other forces, and the cycle repeats. The capillary liquid flow is provided by special microcellular lining on the inner walls of the thermosyphon. In the presence of constant sources of heat and cold, the system can operate successfully without other energy costs. The operation of thermosyphons has a large literature (see, for instance, an overview in [3]). Most publications deal with the conditions of high thermal loads (100 W/m or higher per tube unit length), when the heat transfer is especially active [4]. However, the conditions of low thermal loads (≤10 W/m) applied to the stabilization of permafrost, when the low kinetic energy of the two-phase flow hardly overcomes hydraulic resistance, have received much less attention. The action of the thermosyphons with the simplest construction (as vertical heat pipe) on permafrost is investigated mostly [5, 6]. The operation of some thermosyphon types, such as systems with horizontal evaporator tubes, has been poorly studied theoretically and experimentally. Nevertheless, such devices apply successfully on many constructions of northern areas [7–9].

To bridge the gap, model thermosyphons operated at low thermal loads, which simulate typical permafrost conditions, have been tested in the laboratory [10, 11]. Most thermosyphons used for thermal stabilization of permafrost run only during the cold season and are either vertical tubes or systems with horizontal evaporation tubes (HET). Fluid flow in such systems is commonly driven by gravity, as the use of capillary mechanisms has economic limitations. Where possible, the testing results are formulated as a boundary condition on the evaporator wall at the contact with the ambience, which allows exhaustive characterization of heat exchange between the system and the cooled natural or industrial materials and ensures correct formulation for the respective theoretical problem of external cooling.

2. Vertical thermosyphons

A vertical two-phase free convection system (thermosyphon) consists of relatively short (not over 10–15 m in length) tubes with an inner diameter of 30–50 mm [7, 12–14], based on Long's thermo-valve piling design [15]. The thermosyphons are inserted into the ground to most of their height and have a 1.5–2.0 m stick-up exposed to lower temperatures (**Figure 1**). Their operation commonly stops in summer, and special engineering solutions proceeding from thermal design are required to keep them running all year round or to increase the cooling effect (heat insulation, a certain number of tubes, etc.).

Two-phase free convention thermosyphons have been used to stabilize permafrost under kilometer-long aboveground pipelines (Trans-Alaska system or pipelines in northern, West, and Eastern Siberia), as well as buildings and utility structures in Russian Arctic oil and gas fields [13, 14, 16]. They are intended to cool down the ground to a designed temperature and keep this temperature stable for the whole lifetime of buildings and structures. Thermosyphons are most often installed prior to main construction, when foundations are being prepared. They are almost never used for the energy-consuming freezing of unfrozen parts of the ground, except for few cases of deeply buried systems for thermal stabilization of dam cores or for repairing thaw-related failures in the course of operation [8, 9].

The condenser can be equipped with heat absorption fins which are presumed to enhance the cooling effect, as it has been observed in practice. On the other hand, it follows from the theory [17] that fins should be mounted on the side of the medium with worse thermal properties. Special research is needed, however, to assess the potential efficiency of finning, given that heat transfer in thermosyphons is mainly

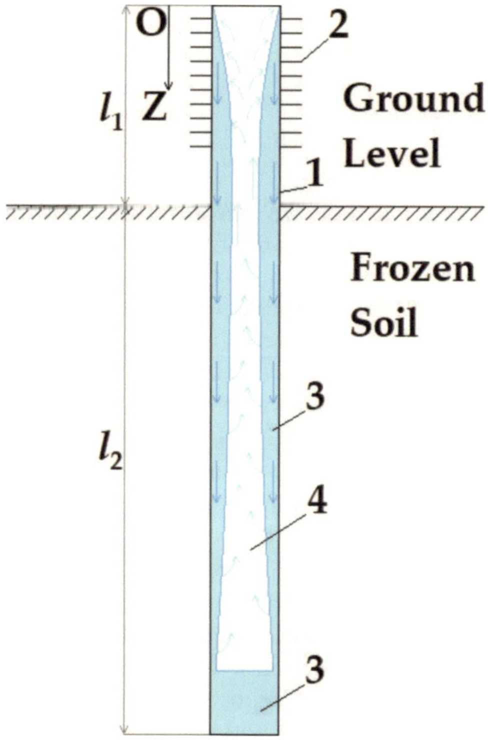

Figure 1.
A sketch of a two-phase thermosyphon: (1) tubular case, (2) fins, and (3 and 4) liquid and vapor phases of cooling fluid.

by convection above the ground and by conduction below the ground which has quite a low thermal conductivity. We have seen no evidence that would prove the thermal performance gain due specifically to fins.

2.1 Condenser

Heat transfer in thermosyphons is modeled jointly for processes inside (fluid phase change and flow) and outside (heat exchange with ground and air) such systems.

The two phases of the working fluid in the tube move as follows [1, 2]. The vapor pressure can be assumed constant along the tube length, which is short in our case, and the hydraulic resistance to vapor flow toward the condenser is very small. Correspondingly, the saturation temperature t_s (related to pressure via the saturation curve) is constant along both evaporator and condenser tubes. The condensed liquid flows along the inner tube as a thin film which thickens up progressively on its way to the evaporator. The film is the thinnest (zero) at the top of the condenser, is the thickest at the evaporator input, and then thins down upon evaporation while moving along the evaporator tube. In the simplest theoretical case, the hydraulic interaction of the two counterflowing phases can be assumed vanishing.

Heat exchange between the condenser and the air can be estimated easily [6, 18]. The total amount of heat Q_k dissipated from the condenser to air per unit time is related to the temperature difference $\Delta t_{ka} = t_k – t_a$ between the condenser wall (t_k) and the incoming air (t_a) as

$$Q_k = \alpha_{tot} \cdot S_k \cdot \Delta t_{sa} \tag{1}$$

$$\alpha_{tot} = \left(\frac{1}{\overline{\alpha}_f} \cdot \frac{S_k}{S_{ik}} + \frac{1}{\alpha_k} \right)^{-1} \tag{2}$$

$$\overline{\alpha}_f = \frac{\lambda_c}{\overline{\delta}}. \tag{3}$$

Note that equations for flat walls are used here and below, because the condensate film is very thin ($\delta << R$, where δ is the film thickness and R is the tube radius), according to observations on transparent physical models. In Eqs. (1)–(3), α_{tot} is the total effective heat loss; S_k is the total area of the condenser outer surface; S_{ik} is the total area of the condenser inner surface; α_k is the heat exchange between the condenser and the air, without the film thermal resistance (found using known relationships via heat exchange of the condenser wall with air [6]); $\overline{\alpha}_f$ is the average heat transfer across the film; $\overline{\delta}$ is the film thickness averaged over the condenser height; and λ_c is the condensate thermal conductivity. The thermal resistance of the condenser wall is neglected as it has quite a high thermal conductivity while the wall is thin.

The average heat transfer across the sinking condensate film $\overline{\alpha}_f$ can be found from known depth dependence of the film thickness $\delta(z)$. At a constant temperature of the condenser wall t_k, the depth-dependent film thickness $\delta(z)$ follows the Nusselt theory [19]. The equation for depth dependence of film thickness, corrected for heat exchange with air, neglecting vapor density (much smaller than liquid density) is given below for a condenser with the axis Oz directed down from the coordinates origin at the condenser top (**Figure 1**):

$$\int_0^z q(z)dz = \frac{\kappa \cdot g \cdot \rho_c}{3 \cdot \nu_c} \cdot \delta^3(z). \tag{4}$$

The heat flux density $q(z)$ at the condenser refers to its inner surface, of the area S_{ik}, and is

$$q(z) = \frac{\Delta t_{sa}}{\delta(z)/\lambda_c + S_{ik}/\alpha_k \cdot S_k}. \tag{5}$$

In Eqs. (4) and (5), ν_c and ρ_c are the kinematic viscosity and density of the condensate, respectively; κ is the heat of vapor-to-liquid phase transition; and g is the gravity acceleration.

Differentiation of both sides in expression (4) along z leads to the equation for $\delta(z)$, at $\delta(0) = 0$:

$$\frac{\nu_c \cdot \Delta t_{sa}}{\kappa \cdot g \cdot \rho_c} = \left(\frac{\delta(z)}{\lambda_c} + \frac{S_{ik}}{\alpha_k \cdot S_k} \right) \cdot \delta^2(z) \cdot \frac{d\delta}{dz}. \tag{6}$$

Integration of this equation leads to a biquadratic algebraic equation with respect to $\delta(z)$:

$$\delta^4 + \frac{4 \cdot \delta^3}{3} \cdot \frac{\lambda_c \cdot S_{ik}}{\alpha_k \cdot S_k} - \frac{4 \cdot \lambda_c \cdot \nu_c \cdot \Delta t_{sa} \cdot z}{\kappa \cdot g \cdot \rho_c} = 0. \tag{7}$$

The solution of Eq. (7) can be written in elementary functions, but it would be cumbersome and inconvenient for further analysis. A more simple relation can be obtained for the small film thickness, with the first term in Eq. (7) much less than the second one:

$$\delta \leq\leq \frac{4}{3} \cdot \frac{\lambda_c \cdot S_{ik}}{\alpha_k \cdot S_k}. \quad (8)$$

As follows from the comparison of relations (8) and (5) (factor *4/3* being insignificant), the film thermal resistance becomes negligible when inequality (8) fulfills. It means that the performance of highly efficient condensers (with large S_c and high α_c, with a large dot product in the denominator of the right-hand side) is limited by the thermal resistance of the film, even commensurate with the resistance of the condenser. Therefore, the fulfillment of inequality (8) places constraints on the design of finned condensers in thermosyphons and in this respect is an economically and technologically important criterion. The respective requirement imposed on the condenser design has to be taken into account in further consideration.

With regard to the formulated requirement, we obtain the equation for the film thickness from Eq. (7), neglecting the first term in the left-hand side:

$$\delta(z) = \left(\frac{3 \cdot \nu_c \cdot \Delta t_{sa} \cdot z}{\kappa \cdot g \cdot \rho_c} \cdot \frac{\alpha_k \cdot S_k}{S_{ik}} \right)^{\frac{1}{3}}. \quad (9)$$

The film is the thickest on the condenser bottom, at $z = l_1$ (**Figure 1**). If inequality (8) fulfills at the maximum film thickness, it obviously fulfills for any thinner film. From relations (8) and (9), it follows that

$$\frac{\alpha_k \cdot S_k}{S_{ik}} \cdot \left(\frac{3^4 \cdot \nu_c \cdot \Delta t_{sa} \cdot l_1}{4^3 \cdot \lambda^3{}_c \cdot \kappa \cdot g \cdot \rho_c} \right)^{\frac{1}{4}} \leq\leq 1. \quad (10)$$

The left-hand side of inequality (10) is dimensionless and depends on the fluid (coolant) type, as well as on the condenser design. The inner and outer surface areas of an unfinned condenser are equal (the wall thickness being neglected), and the first factor in the left-hand side is equal to the heat loss α_a for unfinned (smooth-walled) tubes in the case of cross flow around tubes defined by the known relationship from the specified wind speed [20]. Let the left-hand side of inequality (10) be denoted as Φ with subscripts that refer to different coolants. For R = 16 mm, l_1 = 1.5 m, and the wind speed 3 and 8 m/s, the α_a values are 26.4 and 58.3 W/(m^2·deg), while Φ values are $\Phi_{am} = 8.26 \times 10^{-3}$, 1.83×10^{-2} for ammonia, $\Phi_{fr\text{-}12} = 1.50 \times 10^{-1}$, 3.33×10^{-1} for Freon-12, $\Phi_{fr22} = 1.04 \times 10^{-1}$, 2.31×10^{-1} for Freon-22, $\Phi_{CO_2} = 2.63 \times 10^{-2}$, and 5.79×10^{-2} for CO_2 (at the temperature difference about the maximum Δt_{sa} = 20°C). Thus, inequality (10) fulfills in all cases in the absence of fins (though not perfectly for Freons at high wind speed), and the contribution of film thickness can be neglected for all fluid considered above. On the other hand, fins on the outer surface of the condenser improve external heat exchange, but their enhancement loses sense when the thermal resistance of the film becomes significant.

For a finned condenser, the parameter Φ depends on thickness, spacing, and other parameters of fins (see the fin thickness dependence of Φ in **Figure 2** for different fluids but the same fin parameters and a wind speed of 5 m/s). For the chosen condenser design, inequality (10) fulfills well for all fin thicknesses with ammonia and slightly worse with carbon dioxide. With Freons, only very thin fins can be efficient, but they may be problematic to fabricate. The effect of the thermal resistance of a condensate film is controlled by its thickness and the thermal conductivity of the liquid phase in different coolants. Variations of film thickness at the condenser output as a function of fin thickness for ammonia and Freon-12

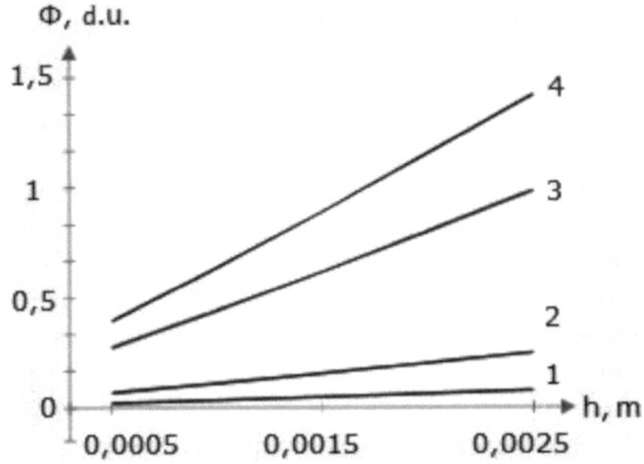

Figure 2.
Behavior of parameter Φ as a function of fin thickness (h) *for ammonia (1),* CO_2 *(2), HCFC (Freon)-22 (3), and HCFC (Freon)-12 (4) fluids, at* R = *16 mm and* l_1 = *1.5 m.*

(**Figure 3**) show that ammonia films are at least 5 times thinner, have about 8 times higher thermal conductivity, and, hence, have about 40 times lower thermal resistance.

As relation (8) fulfills, Eq. (2) for the total heat loss from the condenser becomes simpler:

$$\alpha_{tot} = \alpha_k. \tag{11}$$

2.2 Evaporator

The thickness of the sinking condensate film continuously decreases from the maximum δ_0 at the evaporator input, which coincides with that at the condenser output, and can be estimated by expression (9) at $z = l_1$. The depth-dependent film thickness variations in the evaporator likewise follow the Nusselt theory, applied to evaporation instead of condensation; at a constant temperature on the evaporator wall, t_g is equal to the current temperature of the ambient ground. The previously estimated [21] fluid flow rates inside the system largely exceed the rate of ground

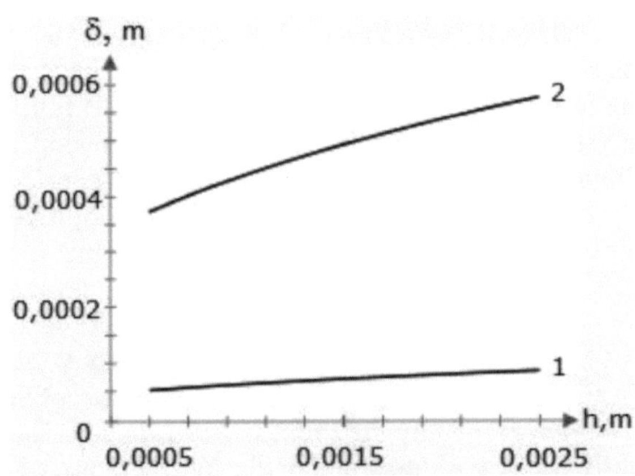

Figure 3.
Film thickness (δ) inside a condenser as a function of fin thickness (h) *for ammonia (1) and HCFC (Freon)-12 (2) fluids.*

temperature redistribution. Therefore, the formation and flow of the condensate film can be considered as a quasi-stationary process at the above conditions. In this case, the heat flux to the evaporator $q(z)$ is

$$q(z) = \frac{\lambda_c \cdot \Delta t_{gs}}{\delta(z)}. \tag{12}$$

The equation of the evaporator heat budget (similar to Eq. (4)) makes basis for the differential equation with respect to film thickness. With the initial condition $\delta = \delta_0$ at $z = l_1$, the equation for $\delta(z)$ becomes

$$\delta^4(z) = \delta_0^4 + \frac{4 \cdot \lambda_c \cdot \Delta t_{gs} \cdot \nu_c \cdot (l_1 - z)}{\kappa \cdot \rho_c \cdot g}; l \geq z \geq l_1, \tag{13}$$

where l is the total tube length. According to expression (13), the nonzero film thickness is limited by some natural depth reasonably corresponding to the maximum evaporator length (see below).

Eq. (13), with averaging along z, can be used to calculate the average heat loss over the evaporator $\overline{\alpha}_{if}$. Note that the parameters t_s and t_g remain unconstrained in all above equations. The saturation temperature t_s is estimated using the total heat budget for the system as a whole; heat coming to the evaporator equals that dissipating to the air from the condenser; $Q_k = Q_f$, where Q_k is defined by relation (1), while the heat from the evaporator is $Q_f = \overline{\alpha}_{if} \cdot S_f \cdot \Delta t_{gs}$ (where the evaporator surface area $S_f = 2 \cdot \pi \cdot R \cdot l_2$ and its length is l_2). Solving the heat budget equation with respect to t_s gives

$$t_s = \frac{S_f \overline{\alpha}_{if} \cdot t_g + S_k \cdot \alpha_k \cdot t_a}{S_f \overline{\alpha}_{if} + S_k \cdot \alpha_k}. \tag{14}$$

With Eqs. (12) and (13), the average heat loss from the evaporator $\overline{\alpha}_{if}$ likewise depends on the temperature t_s and is given by

$$\overline{\alpha}_{if} = \left(\frac{\lambda_c^3 \cdot \kappa \cdot \rho_c \cdot g}{4 \cdot \nu_c \cdot l_2 \cdot \Delta t_{gs}} \right)^{0.25}. \tag{15}$$

To write the equation for t_s, we first define the difference $\Delta t_{gs} = t_g - t_s$ using expression (14), then pick the terms with the difference Δt_{gs}, and obtain the transcendent algebraic equation with respect to this value:

$$\Delta t_{gs} + \frac{3}{4} \cdot \frac{S_f}{\alpha_k \cdot S_k} \cdot \left(\frac{\lambda_c^3 \cdot \kappa \cdot \rho_c \cdot g}{4 \cdot \nu_c \cdot l_2} \right)^{0.75} \cdot \Delta t_{gs}{}^{0.75} = \Delta t_{ga}, \tag{16}$$

which is solved numerically. In the range of ground temperatures t_g from 0°C to t_a, the temperature t_s is close to t_g to thousandth fractions for ammonia and 0.6°C for Freon-12.

The difference Δt_{gs} becomes important for estimating the maximum length l_2^* of the evaporator tube and cannot be neglected. At $\delta = 0$, the maximum tube length is obtained from expression (13) as

$$l_2^* = \delta_0^4 \cdot \frac{\kappa \cdot \rho_c \cdot g}{4 \cdot \lambda_c \cdot \Delta t_{gs} \cdot \nu_c}. \tag{17}$$

Let the heat loss from the condenser in Eq. (9) be $\alpha_k \cdot S_k / S_{ik} \equiv F$ (see its behavior as a function of fin thickness in **Figure 4**) with the values 40, 200, and 400 W/(m^2·deg). Then, l_2^*= 1.7, 6.0, and 11.9 m, respectively, for ammonia and can reach 200 m for Freon-12. This length may be a major advantage of Freon evaporators in thermosyphons used for cooling deep ground.

In principle, evaporator tubes in thermosyphons used in deep ground can be longer than the estimated maximum value, but they should be filled with liquid coolant till the upper limit. In this case, heat transfer should be considered with regard to a shift in the temperature of boiling (in the lower tube section) under the effect of hydrostatic pressure. However, we have no reliable evidence of thermosyphon operation in these conditions.

The reported estimates for the maximum evaporator length require further experimental checks. The issue is especially important because one of us repeatedly observed sporadic ebullition of fluid at the evaporator bottom during laboratory testing of vertical thermosyphons in physical models with transparent walls. The vapor-liquid flow from such ebullition wets the evaporator walls and reaches the condenser, while the evaporator temperature falls abruptly, and moisture condenses on its outer wall surfaces. This effect, which may play a significant role, was demonstrated in a video at TICOP [22], but it is neglected in this consideration. Laboratory testing of thermosyphons has to be continued. The sinking film moves wavelike upon interaction with the down-going flow of vapor [23], but the wavelike flow is stable and even improves the heat exchange to some extent [19].

2.3 External problem for thermal stabilization

The distribution of ground temperatures in the zone of thermal stabilization is estimated as follows. The key issue is to constrain the boundary conditions on the evaporator outer wall which contacts the ground. For this, the saturation temperature (t_s) is expressed from the equation of total heat budget. The heat loss from the condenser Q_k is defined by the same Eq. (11), while the heat coming to the evaporator Q_f from the ground is

$$\alpha_k \cdot S_k \cdot \Delta t_{sa} = S_f \cdot \lambda_f \cdot \left(\frac{\partial t}{\partial r}\right)_R, \tag{18}$$

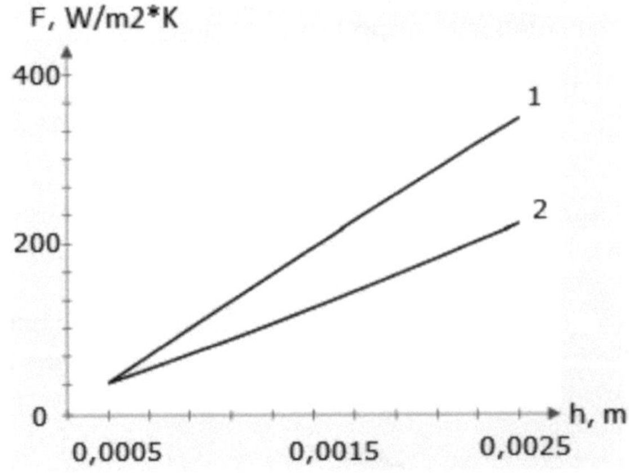

Figure 4.
Behavior of parameter F *as a function of fin thickness* (h)*;* R = 16 *mm.*

where λ_f is the thermal conductivity of frozen ground and t is the ground temperature depending on the radial coordinate r (in the axisymmetric case) and the time τ. With the Eq. (14) for t_s and inequality (8), the sought condition is

$$\frac{\alpha_k \cdot S_k}{S_f} \cdot (t_a - t_g) = -\lambda_f \cdot \left(\frac{\partial t}{\partial r}\right)_R. \quad (19)$$

The same equation was obtained earlier for an unfinned condenser [21], with the only difference in the factor F in the left-hand side of condition (19), before the temperature difference (in the absence of fins, $F_{tot} = \alpha_a l_1/l_2$). Eq. (19) is a third-order boundary condition on the evaporator outer wall ($t_g \equiv t(R)$), which fully characterizes the cooling effect in problems for temperature variation in permafrost stabilized by vertical thermosyphons. The condition allows formulating and solving a large scope of problems concerning the temperature field maintained with any number of vertical thermosyphons in construction. This is the most important result of this part.

3. HET systems

Thermosyphons with horizontal evaporation tubes allow using pad foundations instead of piling in some types of structures, which saves labor costs. These systems have been largely used for permafrost stabilization in terms of works designed mainly by the R&D companies *Fundamentstroiarkos* (direct-flow HET systems) and *Fundamentproekt* (counter-flow systems with inclined evaporator tubes) [8, 24], but no methods are still available for estimating their efficiency. The known theoretical studies characterize the operation of HET systems at high thermal loads of tens or hundreds W/m [4], while the loads in thermal stabilization of permafrost do not exceed 3–5 W/m [11]. As shown by laboratory testing at small thermal loads [10, 11], the two phases of the working fluid have commensurate volumetric contents along the tube length and interact mechanically. This fact does not allow simple description of fluid dynamics and heat exchange in the two-phase flow (e.g., [25–28]). A pure theoretical description of these processes is hardly possible at the time being and requires updating with reference to testing results for specific systems. The original design of HET systems implied unidirectional flow of both phases all along the tubes, which reduces hydraulic resistance to the coolant motion and improves the system performance.

3.1 Methods and results of laboratory testing

The operation of a laboratory model of an HET system is discussed below for a general experiment layout as in **Figures 5** and **6**. The laboratory model was described in detail earlier [10, 11]. The test equipment includes temperature sensors (thermistors), an electronic vacuum meter, a level gauge, and an automatic recorder (**Figure 5**). The condenser is made of metal and placed in a large cooling chamber, while the evaporator consists of glass tubes connected by chemically inert rubber hoses and laid horizontally on wooden pads on the room floor. The transparent glass tubes make visible the flow behavior in different segments of the system, which is important for documenting its operation. Acetone was used as a cooling fluid as its saturation vapor pressure is below the ambient pressure within the applied temperature range (unlike most coolants used in the industry). This ensures cheap, straightforward, and safe handling of the model but requires

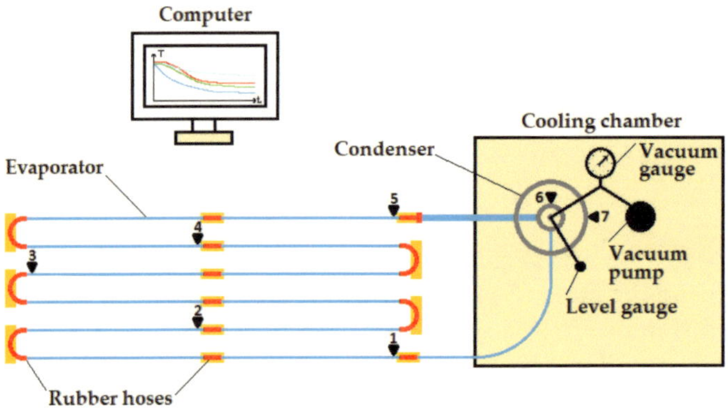

Figure 5.
A sketch of a laboratory test equipment. 1–7 are thermistors.

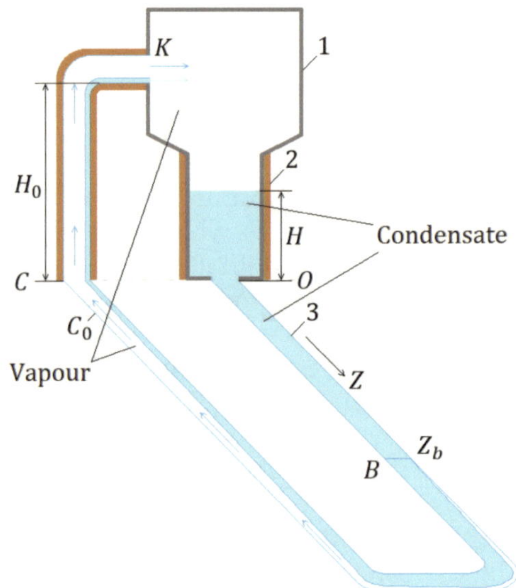

Figure 6.
Structural scheme of the model system (symbols are explained in the text).

pumping out atmospheric gases which are present inside the system and interfere with its work.

The model of the HET system (**Figure 6**) consists of a condenser (1) with a narrower bottom part which accommodates the condensed liquid (2) and an evaporator (3). The vertical evaporator segment leading to the condenser input (segment *CK* in **Figure 6**) is made of a nontransparent insulating material, which masks the fluid flow. The flow behavior within this segment can be judged from examination of the neighbor glass horizontal segment C_0C inside the cooling chamber. When the system operates normally, both fluid phases flow in the same direction, toward the condenser input *K*, while the share of vapor is quite large. The flow of each phase toward point *C* within the main tube length undergoes successively the plug, wave, and projectile regimes.

The system operates normally as long as the tube length-to-diameter (*L/D*) ratio remains ˜500 (in industrial systems, it may reach 5000 or more). As this ratio increases, the operation becomes progressively less stable, while the cooling effect (temperature difference between the evaporator and the room air) reduces. On the other hand, some amount of condensate liquid always flows back into the evaporator during the normal operation. This return flow is visible from outside the cooling chamber [11], except for the nontransparent segment C_0C, and thus forms apparently all along C_0CK (**Figure 7**). The evaporator cools down till the lowest temperature during normal operation, which is recorded by thermistors (**Figure 8**) as distinct from the unstable one (**Figure 9**). The flow of phases within segment C_0C at normal (stable) regime can be either laminar or turbulent (**Figures 10** and **11**).

During the unstable operation, till failure, the liquid phase moves very slowly (visually seeming immobile) within the segment C_0C, while the vapor phase is restricted to bubbles of different sizes (**Figures 12** and **13**) which move relatively slowly toward the condenser input. The cooling effect in this case is often oscillatory and fails to reach the designed temperature at the normal regime (**Figure 9**).

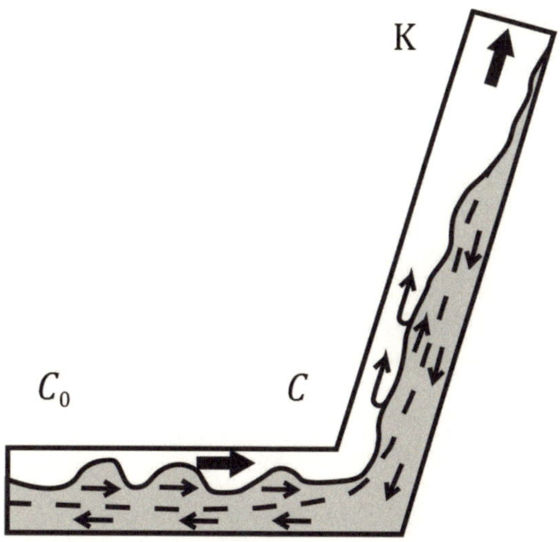

Figure 7.
Two-phase flow near the condenser input (segment C_0CK*): Dark and light shades are condensate and vapor; heavy and thin arrows show vapor and liquid flows, respectively; dash line is the boundary between the two flows.*

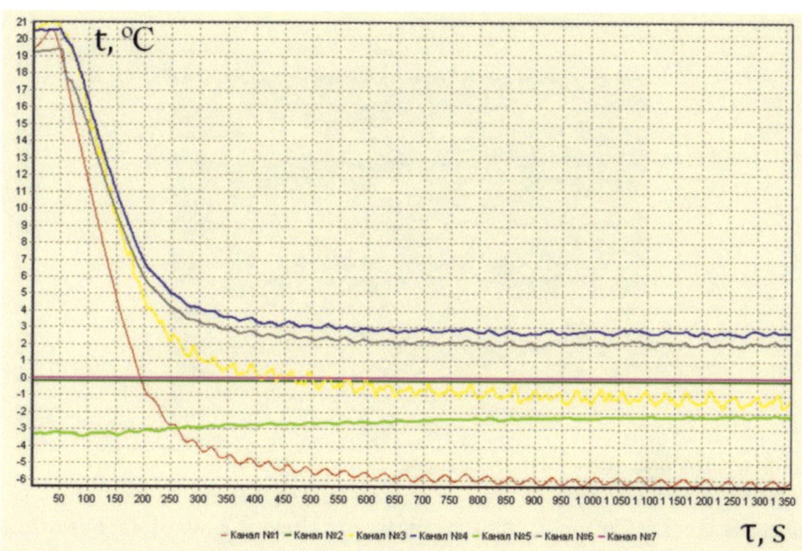

Figure 8.
Thermistor readings along the evaporator tube at stable operation of the system.

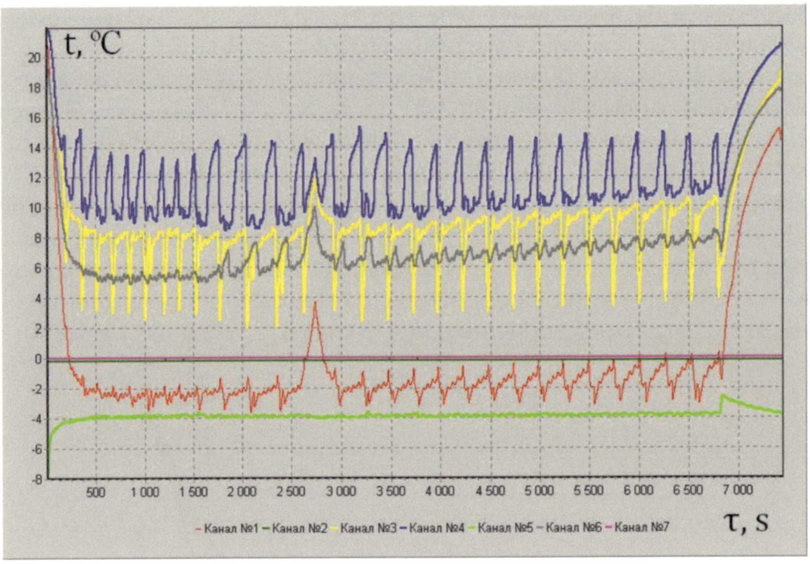

Figure 9.
Thermistor readings along the evaporator tube at unstable operation of the system.

Figure 10.
Laminar flows within segment **C_0C** *at stable operation.*

Figure 11.
Turbulent flows within segment **C_0C** *at stable operation.*

Cooling is the weakest at the smallest amount and slowest flow of vapor within C_0C, that is, the system actually fails. In this case, the condensate level H in the condenser (**Figure 6**) reaches its maximum [11], and constant return flow is visible from outside the evaporator, like during the normal operation. Thus, the return

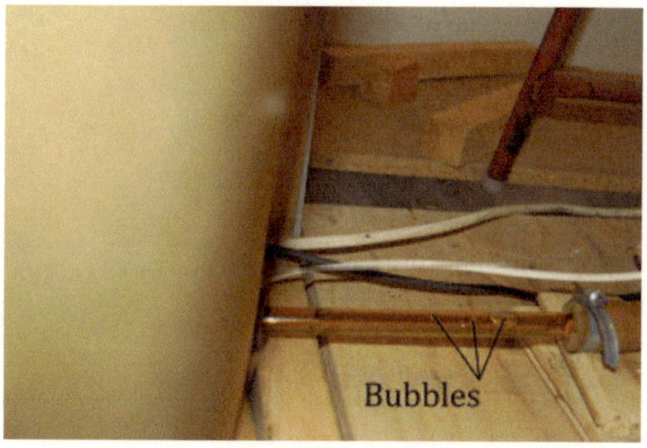

Figure 12.
Small vapor bubbles at unstable operation.

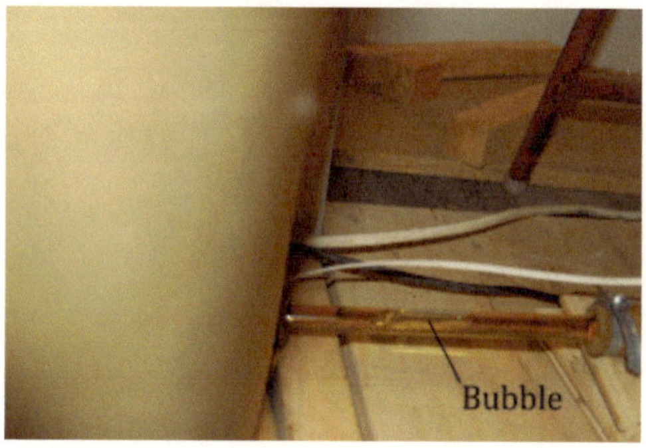

Figure 13.
Big vapor bubbles at unstable operation.

flow along C_0CK is a typical feature of fluid behavior (at least at small thermal loads on the evaporator). Other features (operation instability, slow start, poor cooling effect) become more prominent at longer evaporator tubes (greater L/D ratios) but are minor at short tubes.

The evaporator tubes of full-size HET systems are commonly fully filled with the liquid phase, while the condenser is filled partly, till the specified level H (relative to the evaporator position in the horizontal plane). However, the effect of the fluid volume on the system performance at small thermal loads remains poorly investigated. On the other hand, systems with a lower fluid volume are cheaper and pose lower hazard to environment in case of emergency spilling. The respective laboratory tests were performed also with a partly filled evaporator and an empty condenser ($\beta = L_0/L < 1$, where L_0 is the tube length filled with condensate and L is the total tube length), in order to initially increase the vapor share in the vapor–liquid mixture and thus reduce the hydraulic resistance and improve the system performance. The measured stationary temperature distribution along the tube for different H levels (**Figures 14–17**) demonstrates that the cooling effect reduces notably as the fluid volume increases.

Main testing results can be summarized as follows: (i) there is always a return flow of liquid sinking into the evaporator opposite to the main flow of the two-phase mixture in the segment near the condenser input (segment C_0CK in **Figure 6**) [11]. The presence of this return flow differs the real fluid behavior from the theoretical idea of unidirectional flow of the fluid phases along the whole circulation path (at least for small thermal loads); (ii) at low thermal loads, the systems with return flow develop a slug and plug regime that leads to failure [25]. Specifically, the condensate plugs increase the hydraulic resistance and obstruct the flow to the

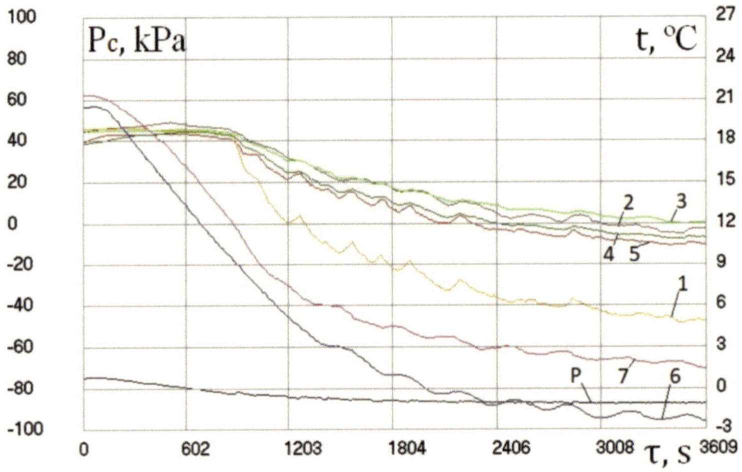

Figure 14.
Temperatures at thermistor points (1–7) and excess pressure (P) during the operation of the model at β = 0.5.

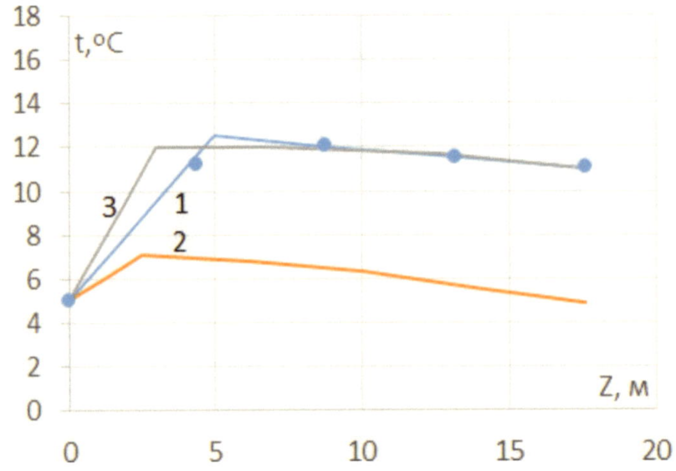

Figure 15.
Measured (1) and calculated (2, 3) temperatures along the tube at β = 0.5, ω = 1 (2), and ω = 50 (3).

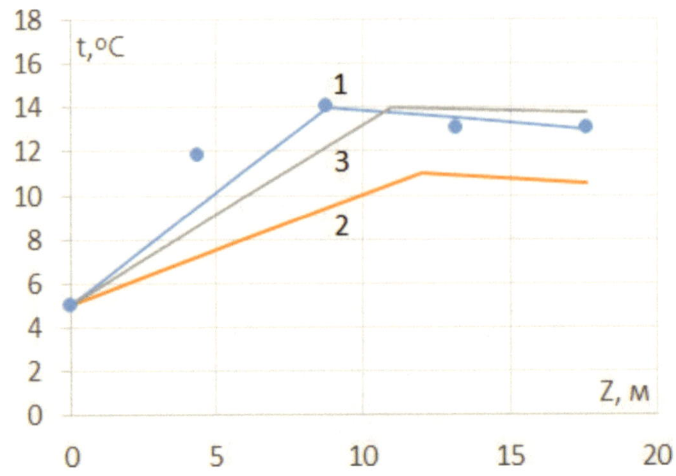

Figure 16.
Measured (1) and calculated (2, 3) temperatures along the tube at β = 1.0, ω = 1 (2), and ω = 70 (3).

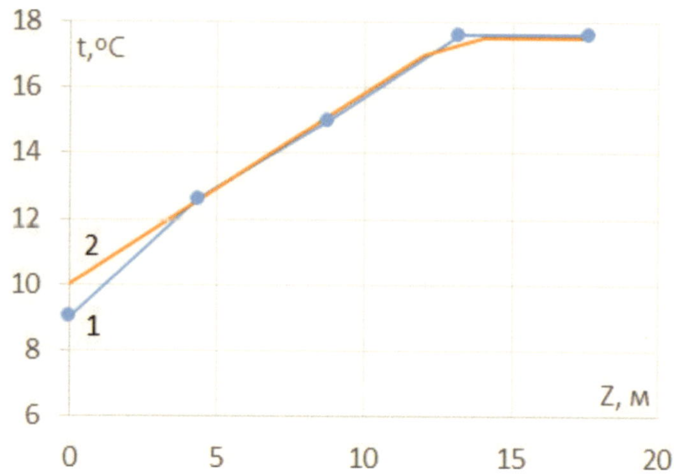

Figure 17.
Measured (1) and calculated (2) temperatures along the tube at standard H level; ω = 1 (2).

condenser within this segment. The flow behavior may change frequently within a single test, which leads to oscillations of the evaporator temperatures (**Figure 9**). This fluid flow behavior at small thermal loads has to be taken into account in calculations for estimating the system performance.

The experimental uncertainty analysis of the experimental data was not done yet. It is planned for the next publications.

3.2 Calculations for thermosyphon design

According to the existing methods of fluid-dynamic calculations [25–28], the state of a two-phase flow in each tube cross section depends on the flow rates of vapor $G_w(z)$ and condensate $G_l(z)$, in kg/s and volumetric vapor content $\alpha_w(z)$, where z is the coordinate along the evaporator tube, with the origin at point O (**Figure 6**). The condensate flow rate is expressed via the total flow rate G: $G_l(z) = G\text{-}G_w(z)$. In the stationary process, G is constant along the flow, but it is unknown a priori and has to be calculated.

The function $\alpha_w(z)$ in the simplest model of Lockhart-Martinelli [26, 27] is defined by $G_w(z)$ and $G_l(z)$. The fluid flow along the tube is split into two parts: heating and two-phase flow. In the heating part, from the point O to the boiling point B at z_b (**Figure 6**), $\alpha_w(z) = 0$, and the coordinate z_b is found from the known solutions for temperature and pressure for the single-phase flow at a given lateral heat transfer and from the condition at which the flow parameters reach the fluid saturation at the given point. In the simplest case of constant heat flux q along the evaporator wall (W/m), the temperature change between O and z_b is

$$t(z) = t_s + q \cdot z/G \cdot c_l, \tag{20}$$

where t_s and c_l are the fluid temperatures at the output of the cooling chamber (point O, **Figure 6**) and the specific heat of the liquid phase, respectively. At the liquid–vapor equilibrium, the saturation pressure P_s correlates with the temperature of the phases t, and the $P_s(t)$ function can be determined empirically as a three-parameter curve, to a sufficient accuracy. Here we use the quadratic equation:

$$P_s(t) = b_1 + b_2 \cdot t + b_3 \cdot t^2, \tag{21}$$

with the coefficients $b_1 = 10^4\ Pa$, $b_2 = 600\ Pa/°C$ and $b_3 = 12.5\ Pa/(°C)^2$ for an acetone fluid. The z-dependent saturation pressure can be expressed as in Eq. 20: $P_s(z) = P_s(t(z))$. The liquid flow within the *OB* segment is laminar, and expression for hydrodynamic pressure $P_h(z)$ is given by

$$P_h(z) = P_k + \rho_l \cdot g \cdot H - \frac{128 \cdot \mu_l \cdot G \cdot (H + z)}{\pi \cdot \rho_l \cdot D^4}, \tag{22}$$

where $P_k = P_s(t_k)$; t_k and P_k are, respectively, the pressure and temperature of saturation vapor inside the condenser; ρ_l and μ_l are the density and dynamic viscosity of the condensate; and D and g are the inner diameter of the tube and the acceleration due to gravity. The position of the boiling point B (z_b) is found from the equation $P_s(z) = P_h(z)$ (the respective equation is omitted). Both t_k and P_k inside the condenser are a minimal value of temperature and pressure for all two-phase segment *BCK*.

In the two-phase flow part, between z_b and the condenser input K (**Figure 6**), the functions $G_w(z)$ and $\alpha_w(z)$ increase monotonically from zero at z_b, while the z dependence is calculated according to the specified conditions of heat transfer on the evaporator wall. In the current model, all incoming heat is spent on fluid evaporation, and $G_w(z)$ is

$$G_w(z) = q(z - z_b)/\kappa. \tag{23}$$

Note that all variables depend on G, H, H_0, and other constructive and external parameters (e.g., on the temperature in the cooling chamber t_a). The pressure gradient within the two-phase flow segment dP/dz is a sum of three components (see [27, 28] for their expressions), $dP/dz = (dP/dz)_f + (dP/dz)_a + (dP/dz)_g$, which refer to friction, flow acceleration, and gravity, respectively. The gravity component disappears within the horizontal tube, and the acceleration component disappears within the vertical tube segment *CK* (with the assumption of ideal insulation of the inlet tube). Integration of the dP/dz equation along the segment between the condenser top (where the pressure is P_k) and condenser input K leads to zero total pressure, and the sought variables can be found by equation

$$\oint \frac{dP(z, G, H)}{dz} dz = 0. \tag{24}$$

In this equation, P depends explicitly on the parameters G and H, which is relevant to the further consideration. The variables t_k and t_s in Eqs. (20) and (22) can be found from the equation for condenser-air heat exchange (with regard to the condenser design and processes inside it) and the integral balance equation for the stationary operation of the system (similar to Eq. (14). In this consideration though, focused on the applicability of the existing calculation methods to small thermal loads, the temperatures are found in a simpler empirical way: from thermistor readings (#1 and #7 in **Figure 5** for t_s and t_k, respectively).

If the condenser design ensures invariable H at any change inside the evaporator during the system operation (i.e., H is a specified and known value in the calculations), Eq. (24), together with known t_s and t_k, is sufficient to find all process parameters. In this case, integration in relation (24) leads to an algebraic equation with respect to G which has a simple numerical solution. Once G has been found, all flow parameters can be found as well: z_b, $G_w(z)$, $\alpha_w(z)$, $P_s(z)$, and the temperature distribution within the two-phase flow part which is related with the saturation pressure $P_s(z)$ as in expression (21). The temperature distribution along the tube should be a continuous function of z, given the way of z_b determination. It is convenient in our case to divide the friction contribution to the integral of Eq. (24) into two components (I_{fL} from the origin to C_0 and I_{fH0} along C_0CK), as high hydraulic resistance within C_0CK is especially important for this consideration.

If the evaporator tube is filled to a part of its length ($\beta \le 1$), the initial fluid level is $H = 0$, while H is set unprompted and is unknown (to be calculated) in operant condition. For this, an additional equation is used, which is a mathematical form of the mass conservation law for fluid in the system, likewise including two unknowns (G and H), assuming equal tube diameters:

$$(1-\beta)\cdot L + (1-\alpha_w(L))\cdot H_0 + H = \int_{z_b}^{L} \alpha_w(z)dz, \tag{25}$$

where z_b and $\alpha_w(z)$ parametrically depend on G and H. Thus, Eqs. (24) and (25) make up a system of algebraic equations with respect to G and H. Solving them jointly yields the unknowns and the temperature along the tube at arbitrary external parameters (temperature in the cooling chamber, heat transfer at the evaporator wall, fluid properties, and system specifications). Cooling of the evaporator relative to its ambience (room air in the case of laboratory testing) is the main parameter of its heat exchange with the host material.

3.3 Calculated and measured data compared

The calculated and measured data can be compared with an example of three laboratory tests at $\beta = 0.5$, $\beta = 1$, and at a standard fluid level ($H = 1.6\ m$). Other experimental conditions are identical: cooling chamber temperature $t_c = -12.5°C$ and room air temperature $t_a = +23°C$, $L = 17.6\ m$, and $D = 0.01\ m$ $(L/D{\sim}2000)$. The thermal load on the evaporator is assumed to be $q = 2.8\ W/m$ proceeding from free convection as the basic mechanism of heat exchange between the evaporator and the ambient air. Time-dependent temperature variations recorded by all thermistors and saturation pressure for the run with $\beta = 0.5$ (**Figure 14**) allow estimating the values $t_k = 1.5$ and $t_s = 5$ (°C) in the case of a quasi-stationary process (similar records for the other runs are not shown). The measured evaporator temperatures for the quasi-stationary process approximated by curve 1 in **Figure 15** differ markedly from those calculated (curve 2) with Eqs. (24) and (25). To improve the fit, the calculated values require correction by a factor of ω (at the integral I_{fH0}), which increases hydraulic resistance within the segment C_0CK. A good fit between measured and calculated temperatures can be achieved with a 50 times higher hydraulic resistance or at ω = 50 (curve 3). In the second run ($\beta = 1.0$, **Figure 16**), the fit improves at ω = 70. This correction for higher hydraulic resistance within the segment C_0CK is valid for partly filled evaporators with the filled-to-total tube length ratios (β) from 0.3 to 1.0. According to preliminary estimates, the correction factor ω in this β range can vary from a few to hundreds of times, depending on variations of other parameters. In the case of standard H level (run 3, **Figure 17**), the fit is good without correction (ω = 1).

4. Conclusions

Improving the performance of vertical thermosyphons by increasing the surface area of fins is limited by the internal thermal resistance of the condensate film that sinks down the condenser inner walls. The thermosyphon-ground heat exchange can be described by the third-order boundary condition, with the respective problem formulation.

Laboratory testing shows that the model of an HET system in the tested design can operate with both partly and fully filled evaporator tubes. The calculations in the former case should additionally include the mass conservation equation for the working fluid. Calculated and measured data fit well with a correction for higher hydraulic resistance in the flow along the vertical evaporator segment leading to the condenser input. The modified calculations remain semiempirical through and require further updating.

Acknowledgements

The authors wish to thank G.M. Dolgikh, director of the *Fundamentstroiarkos* R&D company, and S.N. Okunev, chief engineer of the same company, for the discussions and constructive criticism of this subject matter.

Notes

The work was made in accordance with Program of Fundamental Researches RAN IX. 135.2, Project IX. 135.2.4.

Author details

Yakov B. Gorelik* and Artur H. Khabitov
Earth Cryosphere Institute, Siberian Branch of the Russian Academy of Sciences, Russia, Tyumen

*Address all correspondence to: gorelik@ikz.ru

References

[1] Dunn PD, Reay DA. Heat Pipes. Oxford: Pergamon Press; 1978. p. 334

[2] Pioro IL, Antonenko VA, Pioro LS. Efficient Heat Exchange Systems with Two-Phase Thermosyphons. Kiev: Naukova Dumka; 1991. p. 246 (in Russian)

[3] Vasil'ev LL, Konev SV. Heat pipes. Heat Transfer (Soviet Research). 1974; **6**(1):1-105

[4] Kutepov AM, Sterman AS, Stiushin AG. Fluid Dynamics and Heat Transfer in Vapor Formation. Moscow: Vysshaya Shkola; 1986. p. 448 (in Russian)

[5] Khrustalev LN. Principles of the Geotechnics at Cryolithozone. Moscow: MSU; 2005. p. 554 (in Russian)

[6] Gorelik JB, Seleznev AA. Efficiency of finning in short vertical two-phase thermosyphons for construction on permafrost. Earth's Cryosphere. 2016; **XX**(2):78-89

[7] Dolgikh GM, Okunev SN. Construction and stabilization of cooled foundations in permafrost by *Fundamentstroiarkos* company. In: Proceedings of the International Conference on Theory and Practice of Permafrost: Assessment and Predictions; 29-31 May 2006; Russia. Tyumen: TNGGU; 2006. pp. 228-232

[8] Dolgikh GM, editor. Systems for Thermal Stabilization of Permafrost. A Collection of Papers by Specialists from *Fundamentstroiarkos* R&D Company for 2010–2014. Novosibirsk: Geo; 2014. 218 p (in Russian)

[9] Kutvitskaya NB, Minkin MA. Design of foundations for infrastructure of oil, gas, and condensate field in complex permafrost conditions. Osnovaniya, Fundamenty, Mekhanika Gruntov. 2014;**1**:21-25

[10] Gorelik JB, Gorelik RJ. Laboratory research of work of the natural convectional biphase cooling device with horizontal evaporative part. Kriosfera Zemli. 2011;**XV**(2):34-43

[11] Gorelik JB. Operation of two-phase thermosyphons with horizontal evaporator tubes: Causes of instability. Earth's Cryosphere. 2015;**XIX**(4):81-93

[12] Vasil'ev LL, Vaaz SL. Freezing and Heating of Ground with Thermosyphons. Minsk: Nauka i Tekhnika; 1986. 192 p (in Russian)

[13] Vyalov SS, editor. Ground Temperature Stabilization with Thermosyphons. Yakutsk: IM SO RAN; 1983. 124 p (in Russian)

[14] Makarov VI. Thermosyphons in High-Latitude Construction Engineering. Novosibirsk: Nauka; 1985. 169 p (in Russian)

[15] Long EL. Means for Maintaining Permafrost Foundations. Patents USA, Cl. 165–45; 1964. No. 3, 217, 791

[16] Dolgikh GM, Dolgikh DG, Okunev SN. Freezing of soils: Engineering solutions of *Fundamentstroiarkos* company. In: Proceedings of the International Conference on the Cryosphere of Petroleum Provinces; 26-28 May 2004; Russia. Tyumen: TNGGU; 2004. pp. 56-57

[17] Arnold LV, Mikhailovskiy GA, Seliverstov VM. Engineering Thermodynamics and Heat Transfer. Moscow: Vysshaya Shkola; 1979. p. 445 (in Russian)

[18] Roizen LI, Dulkin IN. Thermal Design of Finned Surfaces. Moscow: Energiya; 1977. p. 255 (in Russian)

[19] Mikheev MA, Mikheeva IM. Fundamentals of Heat Transfer.

Moscow: Energiya; 1973. 320 p (in Russian)

[20] Wong HY. Handbook of Essential Formulae and Data on Heat Transfer for Engineers. London, New York: Longman Group Ltd; 1977. p. 236

[21] Gorelik JB. Modeling ground temperature field around vapor-liquid thermal piles. Problemy Nefti i Gaza Tyumeni. 1980;**47**:58-61

[22] Gorelik JB. Laboratory study of the devices for thermostabilization of frozen ground. Initiative oral reporting. In: Program of 10th International Conference on Permafrost (TICOP); 25-29 June 2012; Russia. Salekhard: Pechatnik; 2012. p. 18

[23] Kapitsa PL. Wave flow of thin liquid layers. Zhurnal Teoreticheskoy Fiziki. 1948;**18**(1):3-18

[24] Minkin MA. Basements and Foundations in Permafrost. Moscow: GASIS; 2005. 213 p (in Russian)

[25] Wallis G. One Dimensional Two-Phase Flow. New York: McGraw-Hill; 1969. p. 440

[26] Butterworth D, Hewitt GF. Two-Phase Flow and Heat Transfer. Oxford University Press (Harwell Series); 1977. 514 p

[27] Crowe CT, editor. Multiphase Flow Handbook. London—New York: Taylor & Francis; 2006. 1128 p

[28] Labuntsov DA, Yagov VV. Mechanics of Two-Phase Systems. Moscow: MEI; 2000. 374 p (in Russian)

Chapter 5

Identification of Heat Exchanger by Neural Network Autoregressive with Exogenous Input Model

Tatang Mulyana

Abstract

This chapter presents the performance of neural network autoregressive with exogenous input (NNARX) model structure and evaluates the training data that provide robust model on fresh data set. The neural network type used is backpropagation neural network also known as multilayer perceptron (MLP). The system under test is a heat exchanger QAD Model BDT921. The real input-output data that collect from the heat exchanger will be used to compare with the model structure. The model was estimated by means of prediction error method with Levenberg-Marquardt algorithm for training neural networks. It is expected that the training data that covers the full operating condition will be the optimum training data. For each data, the model is randomly selected and the selection is based on ARX structure. It was validated by residual analysis and model fit, and validation results are presented and concluded. The simulation results show that the neural network system identification is able to identify good model of the heat exchanger.

Keywords: identification, heat exchanger, neural network, autoregressive with exogenous input, model

1. Introduction

A heat exchanger model QAD BDT921 that is installed in the control laboratory is being used as a model plant to achieve the digital control system design since it is analog in nature. The type of this equipment is a shell and tube heat exchanger, and the application of this model is to generate the heat using steam water. It is the most important equipment in manufacturing and industrial plant in order to maintain and control temperature. In this chapter, more concentration is on the process control and the application of the various methods which are used to develop a mathematical model for the plant. In order to verify the experimental data, a mathematical model is needed to describe the process. Based on the physical and chemical principles, the mathematical models can be derived. However, it is not always possible to develop a model theoretically because the experimental procedures produced by the manufacturer do not provide enough information on how to obtain a mathematical model of the plant. Thus, another useful way is by using system identification to build process models and estimate unknown model parameters. The objective of system identification is to determine system equations from a given input and output time history. There is a few mathematical modeling that can be implemented to the heat exchanger. In this project, the identification of a nonlinear system has been focused on the neural network autoregressive with exogenous input (NNARX). It is one of the neural network system identification classes. It functions using certain syntax from MATLAB command window. Sampling time can be changed due to the system requirement. The heat exchanger is closed loop with controller using proportional integral derivative (PID) control system, while the main objective of this project is to obtain the real input-output data in open-loop system for system identification purposes. Any dynamic response of the process will depend on this controller setting. To open-loop the closed-loop system, the controller must be off by setting the parameter on the temperature controller TIC11. So, the process of heat transfer occurs between hot water from the T12 through the boiler and the cold water inlet (product liquid) at room temperature around 25.1°C. In order to obtain the modeling system of the heat exchanger, the models need mathematical models by using the system identification. Other forms of identification are available and also used in industry such as modeling systems by their physical properties, but it takes a lot of time compared to using system identification since it is only concerned with input and output signals from the system. Neural network processes information in a similar way the human brain does. The network is composed of many elements (neurons) that work in parallel to solve certain problem. The disadvantage is its unpredictable operation because the network figures out how to solve the problem by itself. In the way to complete this chapter, there are a few objectives that have to be achieved. The objectives are:

1. To collect data of heat exchanger QAD BDT921 through experiment design

2. To design NNARX based on multilayer perceptron neural network (MPNN)

3. To simulate NNARX using M-File MATLAB

4. To analyze the simulation resulted based on MSE and LMA

5. To identify mathematical modeling for heat exchanger using network autoregressive with exogenous input (NNARX)

The scopes of this chapter are:

1. Data input is cold water inlet (product liquid) at room temperature around 25.1°C, and the data output is product (water) temperature from range between 22.5 and 55.7°C.

2. NNARX designed are multilayer perceptron (MLP) as structure model, Akaike Final Prediction Error (AFPE) as model training for ARX, Levenberg-Marquardt (LMA) as model training for neural network, and mean square error (MSE) and residual analysis (auto- and cross-correlation) as model validation.

3. To running M-File MATLAB that will be simulation resulted that are data input output on workspace MATLAB, graph data input output, graph ze (model fit) and graph zv (model fit).

4. The analysis based on discrete-time IDOPLY model, loss function, FPE, and MSE.

The expected result of this project is to obtain the data of heat exchanger QAD BDT 921 through experimental design and be able to design the NNARX based on multilayer perceptron neural network (MPNN). It will be simulated using M-File MATLAB. The simulation resulted will be analyzed based on discrete-time IDOPLY model, loss function, FPE, and MSE. Some papers make a study on heat exchangers for different purposes by using various methods, among others, discrete-time model [1, 2], modeling for intelligent control design [3], ARX model [4], digital control design [5–7], PID control design [8–10], parametric and nonparametric identification [11–13], data analysis [14], and NNARX model [15].

2. Heat exchanger

These are diagram of a typical shell and tube heat exchanger, showing the components (A) tubes, (B) tube sheets, (C) shell and shell-side nozzles, and (D) baffles identified in **Figure 1**.

The tubes are the basic component of the shell and tube exchanger, providing the heat transfer surface between one fluid flowing inside the tube and the other fluid flowing across the outside of the tubes. It is most commonly made of copper or steel alloys. The shell is simply the container for the shell-side fluid, and the nozzles are the inlet and exit ports. It normally has a circular cross-section. The heat exchanger is built of a single shelf pass and sixteen tube pass. Inside the shell there are a total of 23 baffles that serve to "mix" to shell-side fluid, and at any cross-section, its temperature tends to be uniformed. The product liquid will enter the shell and the hot water will enter the tube. Flow rate of the product liquid and hot water is the dependent factor of the heat transfer rate. The hot water is heated in the boiler at a constant temperature to supply the heat for heating up the liquid in a heat exchanger. Essentially, the plant consists of two main control loops, the first controls the water levels in the boiler drum and the second controls the liquid temperature. These two control loops are designed as a closed-loop control system. But in this project, the main objective is to obtain the open-loop system. Since the plant is in closed-loop system, the temperature controller (TIC11) needs to be off in order to obtain the open-loop system. So, the process of heat transfer occurs between hot water from the T12 through the boiler and the cold water inlet (product liquid) at room temperature around 25.1°C. As the most common heat transfer industrial equipment is the shell and tube heat exchanger, this equipment is used in QAD Model BDT921 to study temperature control of a heat transfer process. **Figure 2** shows the schematic of heat exchanger QAD BDT921.

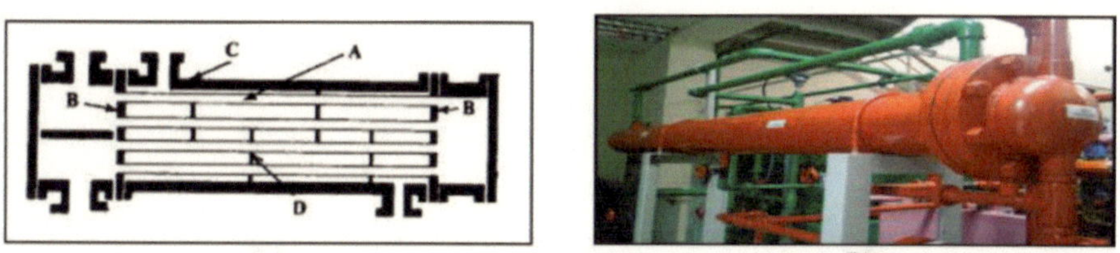

(a) (b)

Figure 1.
(a) Diagram of a typical shell and tube heat exchanger. (b) The heat exchanger QAD BDT921.

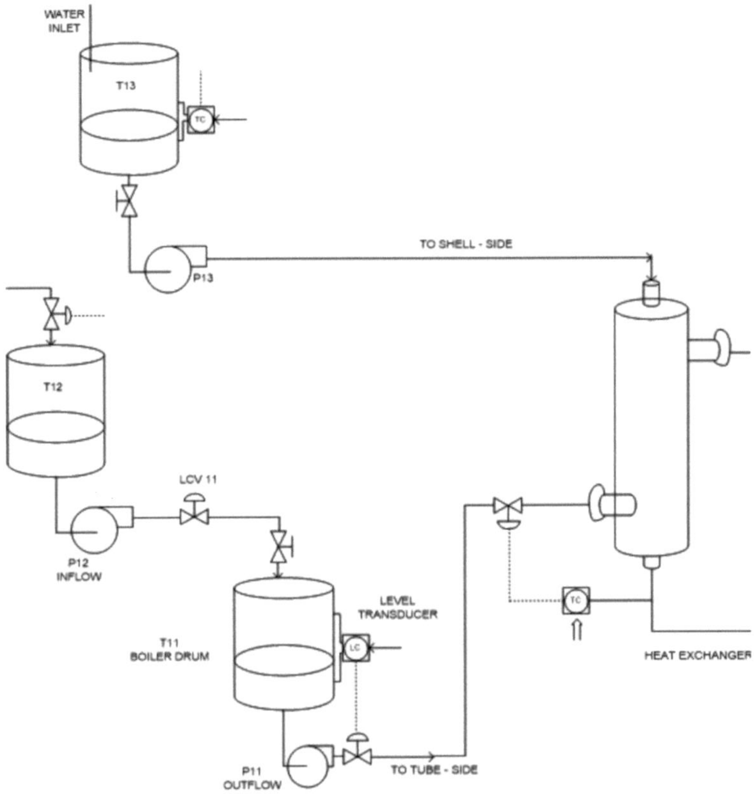

Figure 2.
Schematic diagram of heat exchanger QAD BDT921.

3. Experimental design

The system identification includes the steps such as execute the experiment of heat exchanger and collect input-output data from the process to be identified. The block diagram of open-loop control is shown in **Figure 3**. The data collected by an open-loop controller has the characteristic of not using feedback to determine if its input has achieved the desire goal. The system does not observe the output of the process it is controlling, and it computes its input into system using only the current state and its model of the system. When the data have been collected from the identification experiment, it is often necessary to do some processing on the data set prior to using it for identification. **Figure 4** shows the step of experimental setup in

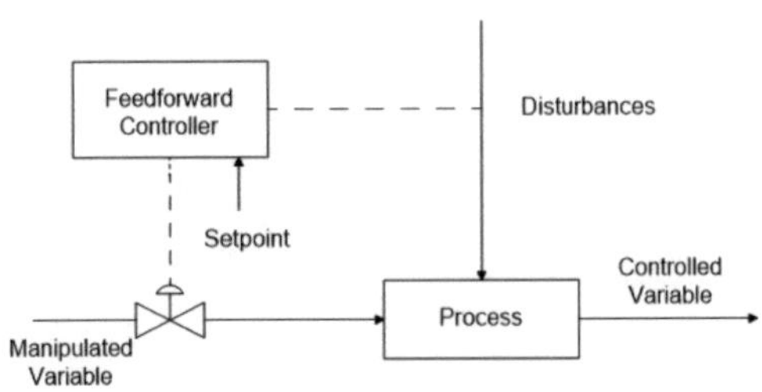

Figure 3.
Feedforward control loop.

DOI: http://dx.doi.org/10.5772/intechopen.83464

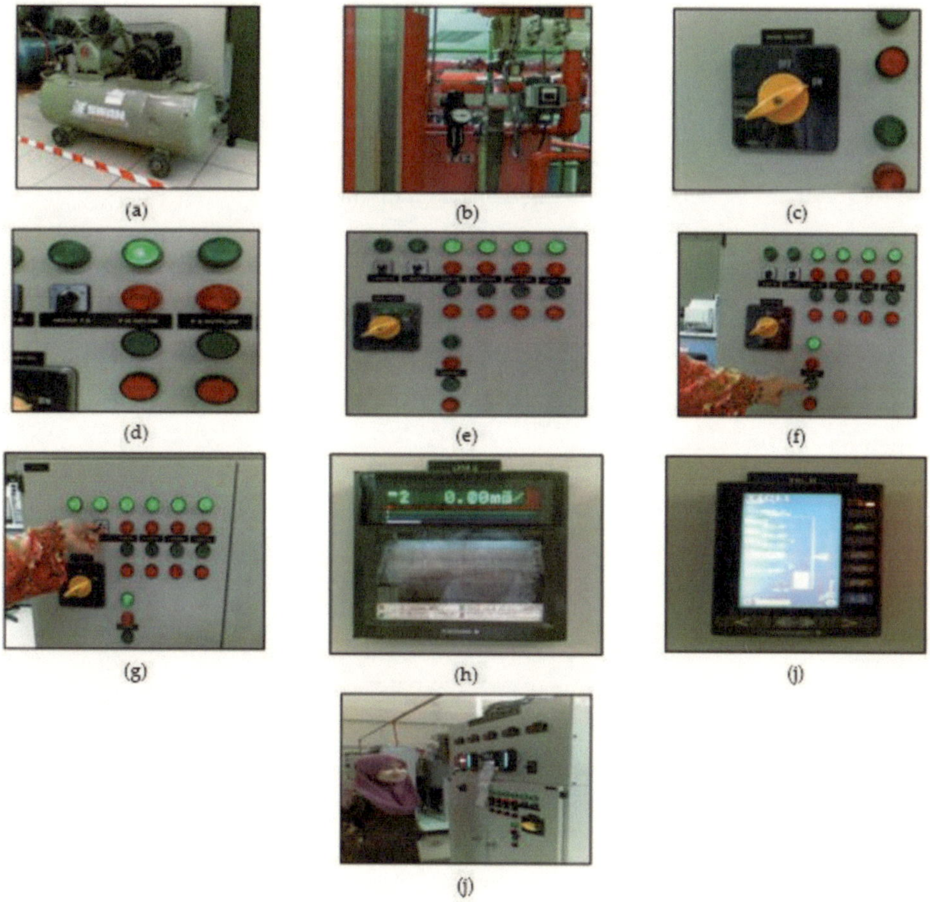

Figure 4.
(a) Start on compressor. (b) Supply AIS. (c) On main switch (415V/3P). (d) On P12 inflow and P11 outflow. (e) P13 and P14 product. (f) On P15 mixing. (g) On heater T12 and T11. (h) Check recorder LFTR1. (i) Temperature controller TIC. (j) Output graph.

obtained open-loop system, and the detail of this experiment, in flowchart form, is shown in **Figure 5**.

The heat exchanger is in closed-loop system, but the main objective of this project is to record the data in open-loop system. To open-loop the closed-loop system, the controller must be in unity state by turning off the temperature sensor. It is off by setting the parameter Pb to 1, Ti to infinity, and Td to 0. The heat exchanger is a shell and tube type used for heat transfer. Preheated tank is used in the plant to supply sufficient hot water. Hot water is heated up in the preheated tank (T12) by a heater starting from zero which needs to be warmed up first. The hot water in the preheated tank T12 is pumped to boiler T11 by water pump P12. The temperature for cold water inlet (product liquid) in tank T13 is at room temperature which is around 25.1°C. It is pumped into the heat exchanger at different flow rates depending on the operator's selection. Operation may select on for water pumps P13 and P14 or either one depending on the flow rate required. The flow indicator FT13 indicates the flow rate of (cold water inlet) product liquid. The product liquid at room temperature is pumped into heat exchanger. The process of heat transfer occurs between hot water from the T12 through the boiler and the cold water inlet (product liquid). The amount of heat transferred in the heat exchanger varies with flow rates of those two liquids. In heat transfer process, the energy

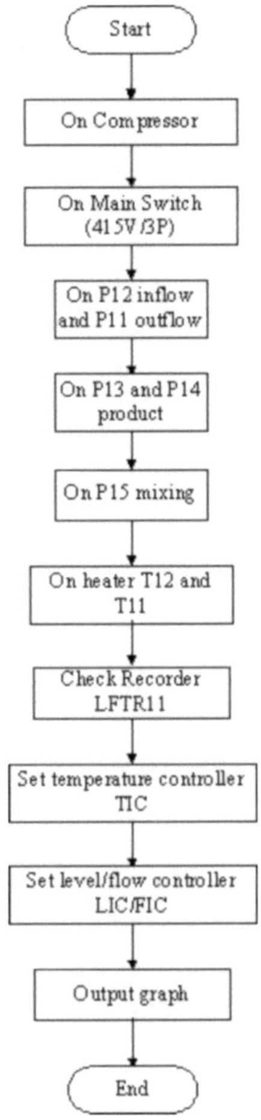

Figure 5.
Flowchart of experimental setup.

stored in one medium as heat capacity is transferred to another medium. The heat exchanger used in the plant is a device where heat is transferred from one fluid, across a tube to another fluid. When two fluids at different temperatures enter the heat exchanger, the temperature of the cold fluid will be increased. Water is used as the medium in the process control training system.

4. NNARX model identification

The most common neural architecture is the multilayer perceptron (MLP). An MLP is a feedforward network also known as a backpropagation neural network built up of perceptron-type neurons, arranged in layers. An MLP has an input layer, one or more hidden layers, and output layer. The feedforward neural network begins with an input layer that may be connected directly to the output layer or a hidden layer. When it is connected to a hidden layer, it can also be connected to another layer whether directly to the output layer or hidden layer. The number is not limited as long as there is at least one output or hidden layer provided. Commonly, there is one hidden layer used in most neural network, neural network that have more than two hidden layers been very rare. **Figure 6** illustrates a typical neural network with a single hidden layer.

In order to develop this system identification, the basic concept of the model structure and estimation method in this project will be discussed. The procedure of modeling and system identification is shown in **Figure 7**.

System identification is the experimental approach to process modeling. It is the process of deriving a mathematical model of a system based on observed input-output data. The procedure to determine a model of a dynamical system from observed input-output data involves three stages which are collecting the real input-output data, choosing a set of candidate model structure and a criterion to select a particular model in the set, and basing on the information in the data. Besides that, in solving this system identification process, a model structure is defined. There are a lot of model structure such as ARX (autoregressive with exogenous input) model, ARMAX (autoregressive moving average exogenous input) model, OE (output error) model, and BJ (Box-Jenkins) model. In this project, the candidate model structure that will be used is NNARX (neural network autoregressive with exogenous input model structure) which is the nonlinear model structures based on neural network. The network structure is defined as a multilayer perceptron.

A linear dynamic system can be represented by Eq. (1):

$$y(t) = q^{-nk}G(q)u(t) + H(q) \tag{1}$$

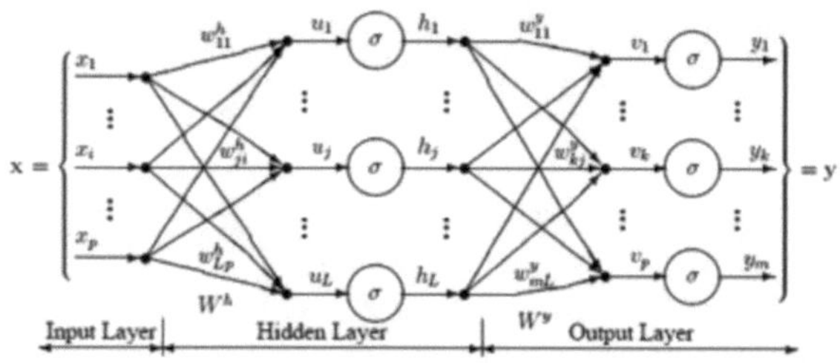

Figure 6.
A typical feedforward neural network (single hidden layer).

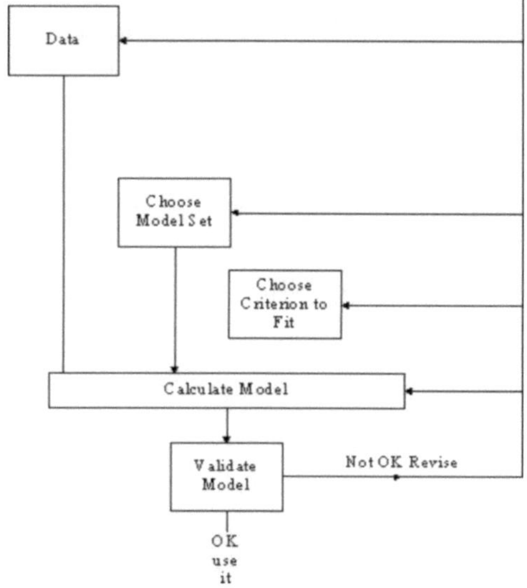

Figure 7.
The procedure of modeling and system identification.

where $q^{-nk}G(q)u(t)$ term refers to noise-free output and $H(q)e(t)$ refers to disturbance term. q is an argument of $G(q)$ and $H(q)$ is the negative shift operator, which is equivalent to q^{-1} represent by q^{-nk} and can be demonstrated by $q^{-1}x(t) = x(t-1)$. nk is the time delay in sampling instant between the process input and the output.

The ARX model structure is given by Eq. (2):

$$y(t) = q^{-nk}\frac{B(q)}{A(q)}u(t) + \frac{1}{A(q)}e(t) \tag{2}$$

where the polynomials A(q) and B(q) are given by Eqs. (3) and (4).

$$A(q) = 1 + a_1q^{-1}+\ldots+a_{na}q^{-na} \tag{3}$$

$$B(q) = b_1q^{-1} + \cdots + b_{nb}q^{-nb} \tag{4}$$

In estimating the nonlinear counterpart of the ARX structure, the neural network can be utilized. According to the research done by M. Norgaard, O. Ravn, N. K. Poulsen, and L. K. Hansen, the multilayer perceptron (MLP) is one of the most popular neural network structures especially in the identification of a nonlinear system. The neural network version of ARX model structure is denoted as ARX (NNARX). Assuming the input delay $nk = 1$, the general NNARX model structure is shown in **Figure 8**.

The input-output relationship of NNARX model structure can be represented by Eq. (5):

$$y(t) = g[\varphi(t) + e(t)] \tag{5}$$

The one-step-ahead prediction (1-SAP) of the NNARX model structure is given by Eq. (6):

$$\acute{y}(t|\theta) = g[\varphi(t), \theta] \tag{6}$$

where $\varphi(t)$ is the regression vector, θ is the parameter vector, g is the function realized by the neural network, $e(t)$ is the noise, $y(t)$ is the system output, and $\acute{y}(t|\theta)$ is the predicted output based on the parameter vector θ.

In this work, the network training algorithm is done by using the prediction error method. The prediction error is given by Eq. (7):

$$\varepsilon(t,\theta) = y(t) - \acute{y}(t|\theta) \tag{7}$$

where $y(t)$ is the observed output, $\acute{y}(t|\theta)$ is the predicted output, and θ is the estimated parameter.

Figure 8.
The general NNARX model structure.

The measurement of the prediction error is often represented by a function known as the loss function. Its general representation can be written by Eq. (8):

$$V_N = \frac{1}{2N}\sum_{t=1}^{N} \varepsilon^2 (t, \theta) \tag{8}$$

where Z^N refers to the training data set. In this project, the network training algorithms used Levenberg-Marquardt training and allow up to 100 epochs to train the network.

The value for this Akaike Final Prediction Error using the Eq. (9).

$$\text{FPE} = \frac{1+\text{n}/\text{N}}{1+\text{n}/\text{N}} . V_N \tag{9}$$

where n is the total number of estimated parameters and N is the length of the data record. This criterion reflecting the prediction error variance says that in collection of different models, the one with the smallest value of FPE should be chosen. Once a model structure has been identified, it is important to validate the model using a data set. Model validation is needed to verify that the identified model fulfills the modeling requirement according to the subjective and objective of good model approximation.

The model validation algorithm used in this identification is mean square error method and residual analysis. The average squared error is shown by Eq. (10):

$$\text{MSE} = \frac{1}{\text{N}}\sum_{t=1}^{\text{N}} \varepsilon^2(t) \tag{10}$$

This is a measure in a single positive number of how well the model output fits the measured data.

It is good to check the autocorrelation of the residual analysis by Eq. (11) as it is assumed to be a white noise sequence:

$$R_\varepsilon^N = \frac{1}{N}\sum_{t=1}^{Nn} \varepsilon(t)\, \varepsilon(1-\tau) \tag{11}$$

If these numbers are not small for $\tau \neq 0$, then part of ε could have been predicted from past data, and so this is a sign of deficiency in the model. Similarly the residuals should not be correlated with the input, so it is also good to check the cross-correlation of the residuals and the input by Eq. (12):

$$R_{\varepsilon u}^N = \frac{1}{N}\sum_{t=1}^{Nn} \varepsilon(t)\, u(t-\tau) \tag{12}$$

5. Result and analysis

In this part, the result for data collection and data prefiltering or data preprocessing will be displayed. The 1000 sample data will be collected by sampling the output graph of the heat exchanger QAD BDT921. **Figure 9** shows an example part of data on the graph.

In this chapter, the important part is data collection. This is because, in this project, the model will be tested based on the real data obtained from the heat exchanger process. In this experiment, the input and output data are temperature. About 1000 sample data were collected from the graph. **Table 1** shows example of 20 data for input and output.

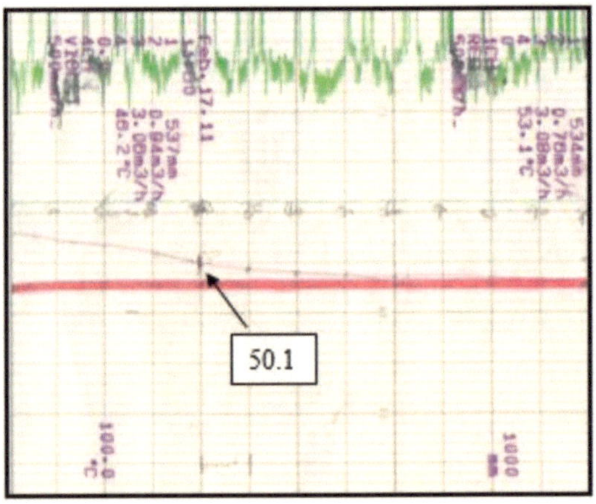

Figure 9.
Example of data on the graph.

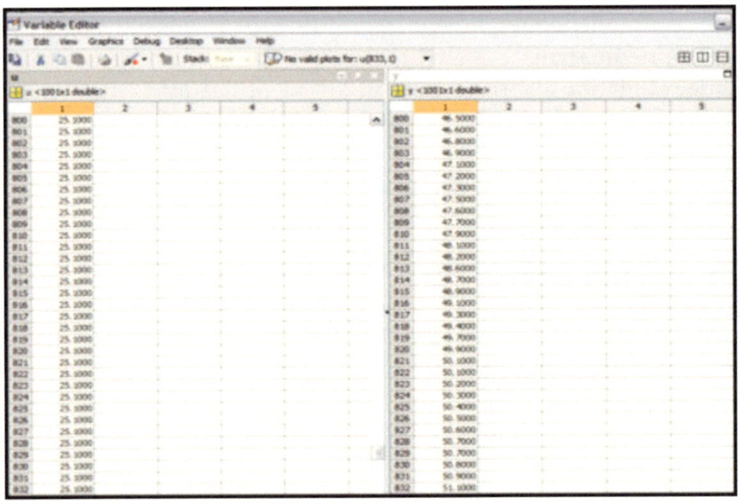

Variable Editor

u <1001x1 double> / y <1001x1 double>

	u (1)	y (1)
800	25.1000	46.5000
801	25.1000	46.6000
802	25.1000	46.8000
803	25.1000	46.9000
804	25.1000	47.1000
805	25.1000	47.2000
806	25.1000	47.3000
807	25.1000	47.5000
808	25.1000	47.6000
809	25.1000	47.7000
810	25.1000	47.9000
811	25.1000	48.1000
812	25.1000	48.2000
813	25.1000	48.6000
814	25.1000	48.7000
815	25.1000	48.9000
816	25.1000	49.1000
817	25.1000	49.3000
818	25.1000	49.4000
819	25.1000	49.7000
820	25.1000	49.9000
821	25.1000	50.1000
822	25.1000	50.1000
823	25.1000	50.2000
824	25.1000	50.3000
825	25.1000	50.4000
826	25.1000	50.5000
827	25.1000	50.6000
828	25.1000	50.7000
829	25.1000	50.7000
830	25.1000	50.8000
831	25.1000	50.9000
832	25.1000	51.1000

Table 1.
The 20 data for input and output of the heat exchanger QAD BDT921.

After data collection has been done, the next step is data prefiltering or data preprocessing. The input and output data obtained from the experiment are plotted using MATLAB software. **Figure 10** shows the results for input and output data (**Figure 11**).

Several types of parameters are used to create system identification with M-file. The parameters are listed as time sampling, orders, and layer size for neural network. Those parameters [Ts] [na nb nk] [layer size] have been analyzed with M-file and also the effect of this parameters. The values (0.072, 0.72) for the number of units in time sampling with the same orders [1 1 1] were tested whether the model fit improves for the data estimation and validation. The effect of time sampling on model fit analyzed with M-file is shown in **Figures 12** and **13**.

One of the important parameters is orders. For orders, the value of [na nb nk] will be set randomly. In this case, the examples of selected orders are [1 1 1] and [1 2 1] with the same layer sizes [2 1] will be tested whether the model fit improves for the estimation and validation data. **Figures 14** and **15** show the effect of orders [1 1 1] on model fit for the data estimation and validation.

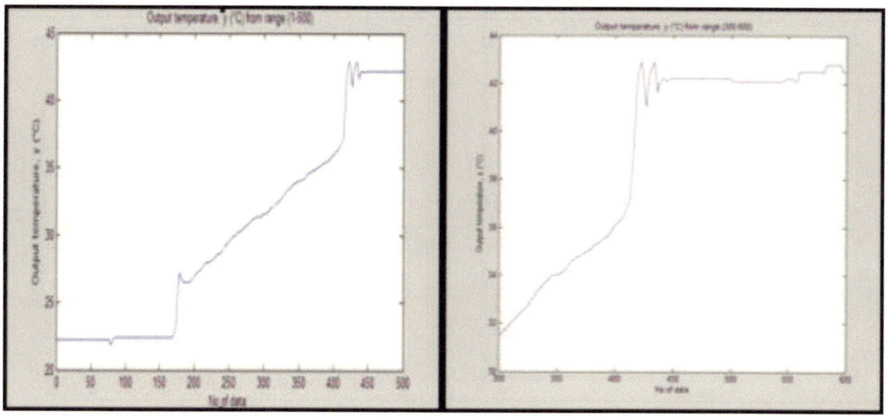

Figure 10.
Output temperature from range (1–500) and (300–600).

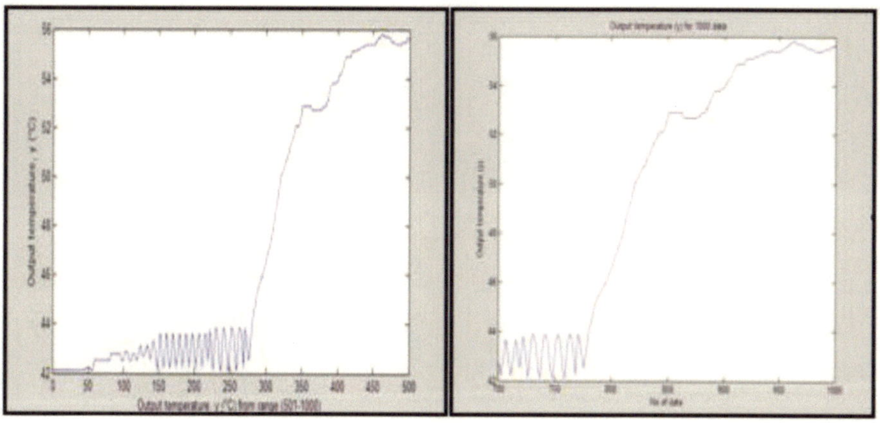

Figure 11.
Output temperature from range (501–1000) and (700–1000).

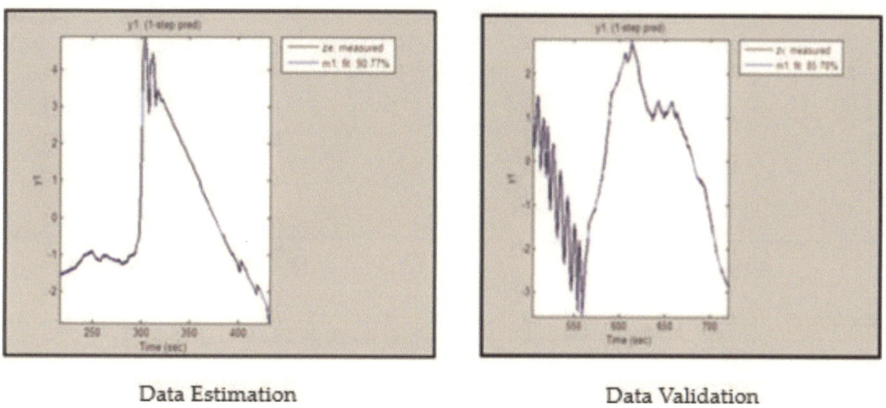

Figure 12.
1-SAP by the model against the data estimation and validation for Ts = 0.072s and [0.072][1 1 1][2 1].

In this case, the neurons' values (2, 4, 8) for the number of units in the hidden layer will be tested to see if the model fit improves for the estimation and validation data. The model used one past value of the input and one past value of the output

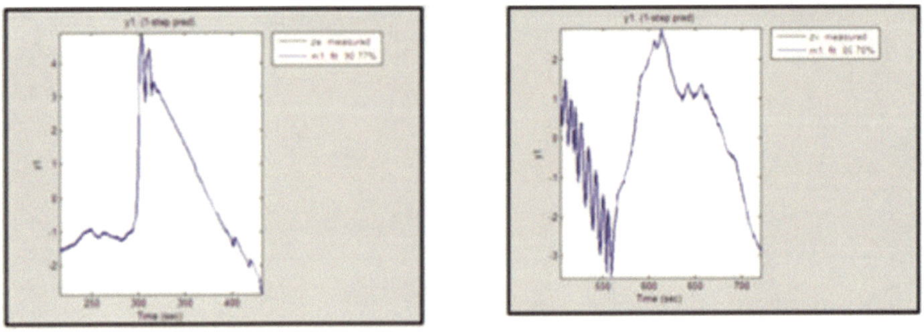

Data Estimation　　　　Data Validation

Figure 13.
1-SAP by the model against the data estimation and validation for Ts = 0.72s and [0.072][1 1 1][2 1].

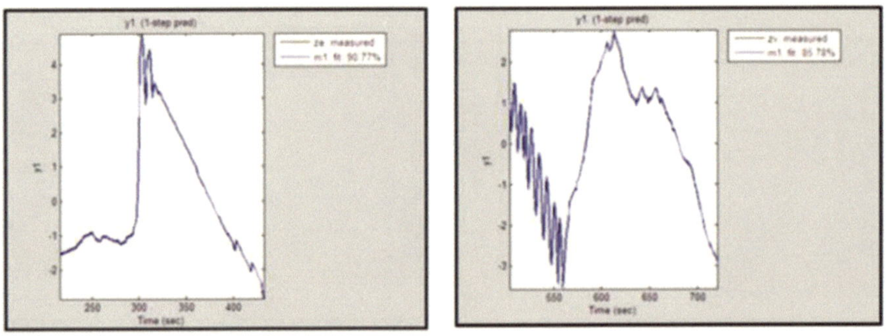

Data Estimation　　　　Data Validation

Figure 14.
1-SAP by the model against the data estimation and validation for [0.072][1 1 1][2 1] at [1 1 1].

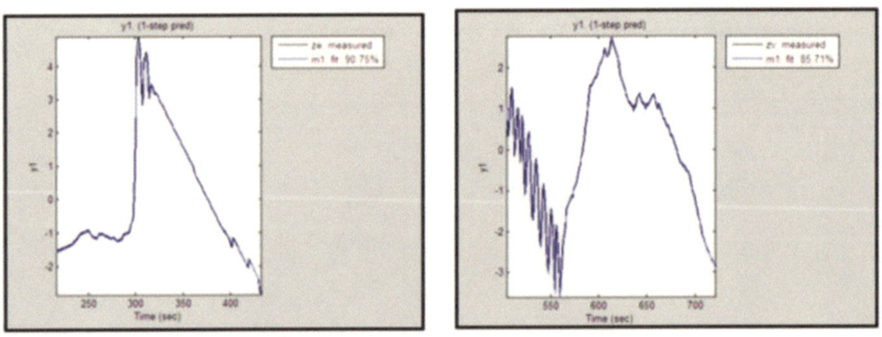

Data Estimation　　　　Data Validation

Figure 15.
1-SAP by the model against the data estimation and validation for [0.072][1 1 1][2 1] at [1 2 1].

[1 1 1]. **Figure 16** shows the one-step-ahead prediction results of optimized NNARX model for estimation data at orders [1 1 1] with 2, 4, and 8 hidden units.

Figure 17 shows the one-step-ahead prediction results of optimized NNARX model for validation data at orders [1 1 1] with 2, 4, and 8 hidden units.

Table 2 shows the effect time sampling (Ts) on model fit for NNARX model based on **Figures 12** and **13**. At 0.72 second, the model fit for estimation data (ze) is 90.96% and for validation data (zv) is 87.64%, while at 7.2 second the model fit is

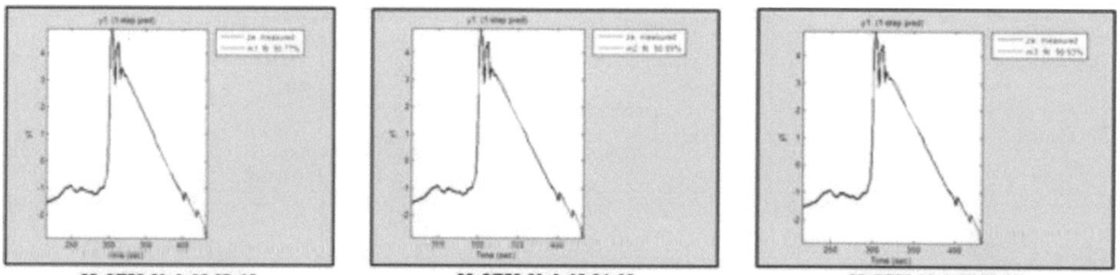

[0.072] [1 1 1] [2 1]
2 hidden units

[0.072] [1 1 1] [4 1]
4 hidden units

[0.072] [1 1 1] [8 1]
8 hidden units

Figure 16.
Estimation plot for model with 2, 4, and 8 hidden units for [0.072] [1 1 1].

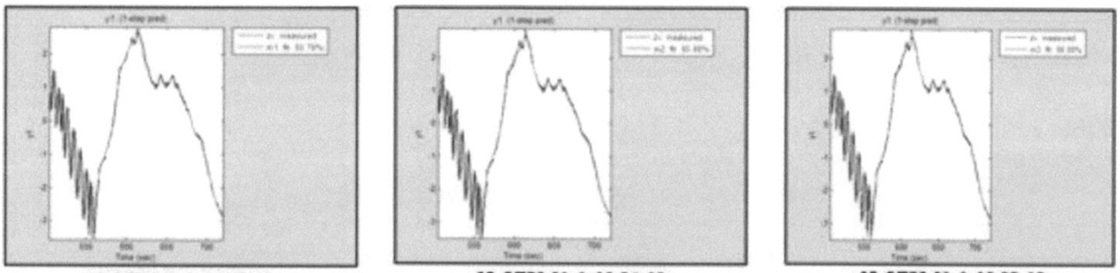

[0.072] [1 1 1] [2 1]
2 hidden units

[0.072] [1 1 1] [4 1]
4 hidden units

[0.072] [1 1 1] [8 1]
8 hidden units

Figure 17.
Validation plot for model with 2, 4, and 8 hidden units for [0.072] [1 1 1].

Ts (second)	Order [na nb nk]	Layer	Model fit (%)	
			ze	zv
0.072	[1 1 1]	[2 1]	90.96	87.64
0.72	[1 1 1]	[2 1]	90.96	87.64

Table 2.
Effect of time sampling on model fit.

also the same for estimation data (ze) and validation data (zv). These results clearly indicate that sampling time did not effect on the model fit. It is the number of samples per second taken from a continuous signal to make it discrete, and holding time is the time between two samples.

Based on **Figures 14** and **15**, the best orders turn out to be na = 1, nb = 2, and nk = 2. The higher order [1 2 2] ARX model is able to reproduce the validation data best, but the differences between the orders are really minor. **Table 3** shows the comparisons between ze and zv at sampling time, Ts = 7.2 second, and layer size [2 1].

The effect of the number of units in the hidden layer on model fit for NNARX model is shown in **Table 4**, where the fit for the validation data with different numbers of units in the hidden layer more accurate. Validation plot for model with two neurons in the hidden layer has a model fit of about 85.78%, while for model with four neurons the model fit is 85.88%, and for model with eight neurons the model fit is 86.06%. Out of the three models found, the one with eight hidden units provides the best fit for the validation data. In this case, the model fit for each hidden units did not improve for estimation data.

The MATLAB result for least square estimates of ARX model at orders [1 1 1] and [3 2 2] is shown in **Figure 18**. It could be seen in **Figure 18** that the results for discrete-time IDPOLY are $A(q) = 1 - 0.9969q^{-1}$ and $B(q) = 1.717e011q^{-1}$. The loss

Ts (second)	Order [na nb nk]	Layer	Model fit (%)	
			ze	zv
0.072	[1 1 1]	[2 1]	90.77	85.78
0.72	[1 2 2]	[2 1]	90.76	85.81

Table 3.
Effect of orders on model fit.

Orders [nn]			Model fit %					
			Hidden layer [2 1]		Hidden layer [4 1]		Hidden layer [8 1]	
na	nb	nk	ze	zv	ze	zv	ze	zv
1	1	1	90.77	85.78	90.81	85.88	90.77	86.06

Table 4.
Effect of the number of units in the hidden layer on model fit.

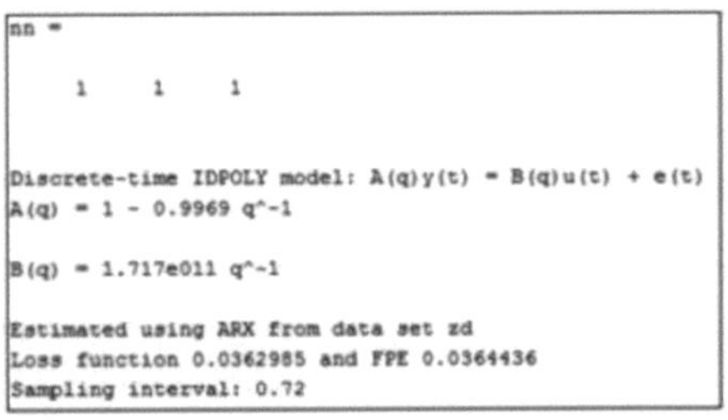

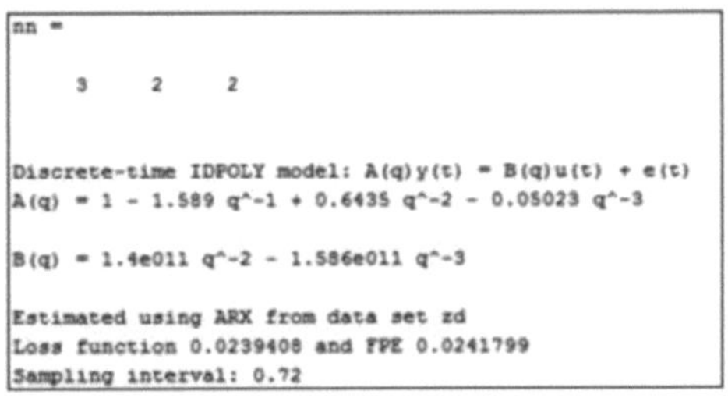

Figure 18.
Least square estimates of ARX model with orders [1 1 1] and [3 2 2].

function for this model is equal to 0.0362985, and the final prediction error is equal to 0.0364436. The value for this Akaike Final Prediction Error has been calculated by MATLAB using the equation of FPE = V*(1 + d/N)/(1 − d/N) where V is the loss function of the model, d is the number of estimated parameters, and N is the number of estimation data. The best selected model is at orders [3 2 2] because it has smallest value loss function and FPE. **Figure 18** shows that the loss function at these orders is 0.0239408 while for the FPE is 0.0241799, and the result for the discrete-time IDPOLY model is $A(q) = 1 - 1.589q^{-1} + 0.6435q^{-2} - 0.05023q^{-3}$ and $B(q) = 1.4e011q^{-2} - 1.586e011q^{-3}$. Although this gives the best result, the suitable model will be decided until the validation process will be done.

Table 5 shows the results for estimated parameter using ARX model from data (zd) at several orders consisting of IDPOLY model, loss function, and FPE. From the table, it can be seen that when the model order selection is low, then the model will have high loss function and FPE. The best estimated parameter is obtained from the smallest value of loss function and FPE. The best selected model is also at orders [3 2 2] and layer size [8 1] because the estimated results using NNARX with nonlinearity estimator (neuralnet) for loss function have a smallest value at these orders. **Table 5** shows the estimated results using ARX from data set zd for IDPOLY model, loss function, and FPE for various orders.

Results for nonlinear regressors are estimated using NNARX with nonlinearity estimator (neuralnet) presented in **Table 6**.

The mean square error was used to evaluate the performance of each neural network. **Table 7** shows the mean square error (MSE) performance for training.

Order [na nb nk]	IDPOLY model	Loss function	FPE
1 1 1	$A(q) = 1 - 0.9969q^{-1}$ $B(q) = 1.717e011q^{-1}$	0.0362985	0.0364436
1 2 1	$A(q) = 1 - 0.9969q^{-1}$ $B(q) = -2.066e011q^{-1} + 2.72e011q^{-2}$	0.0357804	0.0359949
1 2 2	$A(q) = 1 - 0.9969q^{-1}$ $B(q) = 5.001e010q^{-2} - 3.074e010q^{-3}$	0.0354304	0.0356427
2 1 1	$A(q) = 1 - 1.561q^{-1} + 0.5661q^{-2}$ $B(q) = 4.573e010q^{-1}$	0.0242571	0.0244025
2 1 2	$A(q) = 1 - 1.561q^{-1} + 0.5655q^{-2}$ $B(q) = 2.155e010q^{-2}$	0.0242772	0.0244227
2 2 1	$A(q) = 1 - 1.566q^{-1} + 0.5707q^{-2}$ $B(q) = 1.293e011q^{-1} - 7.543e010q^{-2}$	0.0242369	0.0244306
2 2 2	$A(q) = 1 - 1.56q^{-1} + 0.5651q^{-2}$ $B(q) = 1.637e011q^{-2} - 1.814e011q^{-3}$	0.0240013	0.0241931
3 1 1	$A(q) = 1 - 1.589q^{-1} + 0.6556q^{-2} - 0.06161q^{-3}$ $B(q) = 1.359e010q^{-1}$	0.0240461	0.0242383
3 1 2	$A(q) = 1 - 1.59q^{-1} + 0.656q^{-2} - 0.06091q^{-3}$ $B(q) = -3.576e010q^{-2}$	0.0240302	0.0242223
3 2 2	$A(q) = 1 - 1.589q^{-1} + 0.6435q^{-2} - 0.05023q^{-3}$ $B(q) = 1.4e011q^{-2} - 1.586e011q^{-3}$	0.0239408	0.0241799

Table 5.
Results for estimated using ARX from data set zd.

Order [na nb nk]	Regressors	Loss function		
		Layer sizes [2 1]	Layer sizes [4 1]	Layer sizes [8 1]
1 1 1	y1(t − 1) u1(t − 1)	0.029315	0.0269315	0.28331
1 2 1	y1(t − 1) u1(t − 1) u1(t − 2)	0.029486	0.028948	0.020861
1 2 2	y1(t − 1) u1(t − 1) u1(t − 3)	0.029567	0.029033	0.027956
2 1 1	y1(t − 1) y1(t − 2) u1(t − 1)	0.0192380	0.010579	0.0077851
2 1 2	y1(t − 1) y1(t − 2) u1(t − 2)	0.0192380	0.010579	0.0077851
2 2 1	y1(t − 1) y1(t − 2) u1(t − 1) u1(t − 2)	0.0212208	0.011918	0.0049208
2 2 2	y1(t − 1) y1(t − 2) u1(t − 2) u1(t − 3)	0.021253	0.011954	0.0048163
3 1 1	y1(t − 1) y1(t − 2)	0.016355	0.0065896	0.0046813

Order [na nb nk]	**Regressors**	**Loss function**		
		Layer sizes [2 1]	**Layer sizes [4 1]**	**Layer sizes [8 1]**
	y1(t − 3) u1(t − 1)			
3 1 2	y1(t − 1) y1(t − 2) y1(t − 3) u1(t − 2)	0.016355	0.0065896	0.0046813
3 2 2	y1(t − 1) y1(t − 2) y1(t − 3) u1(t − 2) u1(t − 3)	0.016355	0.0063295	0.0045067

Table 6.
Results for nonlinear regressors estimated using NNARX with nonlinearity estimator (neuralnet).

Order [na nb nk]	**Mean square error (MSE)**		
	Layer sizes [2 1]	**Layer sizes [4 1]**	**Layer sizes [8 1]**
1 1 1	0.0292	0.0285	0.0292
1 2 1	0.0293	0.0293	0.0266
1 2 2	0.0293	0.0293	0.0285
2 1 1	0.0105	0.0105	0.0101
2 1 2	0.0105	0.0105	0.0101
2 2 1	0.0211	0.0118	0.0063
2 2 2	0.0210	0.0118	0.0049
3 1 1	0.0162	0.007	0.0047
3 1 2	0.0162	0.007	0.0047
3 2 2	0.0162	0.0064	0.0045

Table 7.
MSE performance for model structures with nonlinearity estimator (neural network).

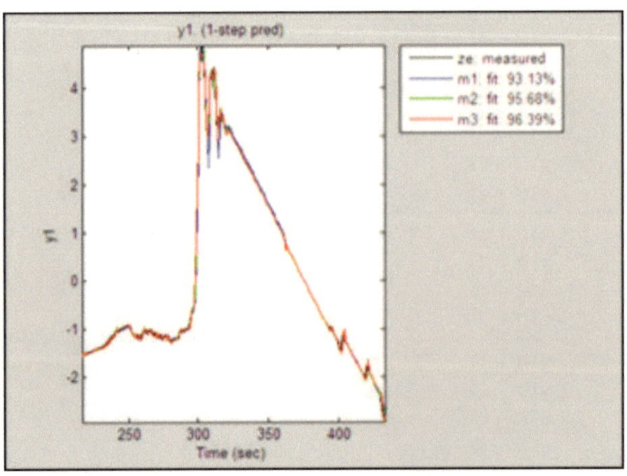

Figure 19.
1-SAP by the model against the estimation data of layer sizes (2, 4, and 8).

From observation, the model has smallest MSE when the model order selection is high. The best model which has the smallest value of MSE of about 0.0045 has order [3 2 2] and at layer size [8 1].

Figure 19 shows the comparison between the system and the model output at order [3 2 2]. The output model trend is similar with the output system. The model fit is resultant of three types of layer sizes such as 2, 4, and 8 in the hidden layer. It can be observed that the best fit for estimation data is at layer (8 1). The model fit is 96.39%.

Based on all the results from the simulation graph, it generally can be said that the model fit is satisfactory in all conditions of training and testing with percentage fit between 90 and 97%. **Table 8** shows the orders consisting of all combinations na, nb, and nk in the range 1–3 and hidden layer effect of the model fit for data estimation. The best identified model has na = 3, nb = 2, and nk = 2 at layer size [8 1] where the m1 is 96.39%. The medium model turns out to be na = 3, nb = 1, and

Time	Orders [nn]			Model fit %		
				Layer size [2 1]	Layer size [4 1]	Layer size [8 1]
	na	nb	nk	m1	m2	m3
Ts = 0.072	1	1	1	90.77	90.81	90.77
	1	2	1	90.75	90.76	91.19
	1	2	2	90.76	90.76	90.87
	2	1	1	95.23	94.46	94.58
	2	1	2	92.53	94.46	94.58
	2	2	1	92.16	94.12	95.71
	2	2	2	92.16	94.12	96.21
	3	1	1	93.13	95.47	96.3
	3	1	2	93.13	95.47	96.3
	3	2	2	93.13	95.68	96.39

Table 8.
Comparisons between several different orders and layer sizes, based on the fit between estimation data and simulated output.

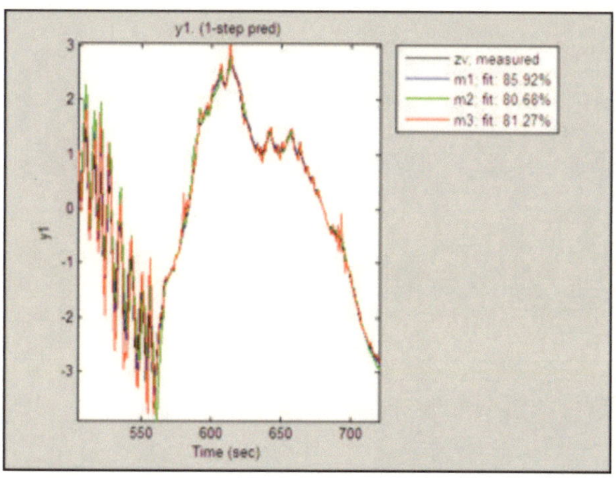

Figure 20.
1-SAP by the model against the validation data of layer sizes (2, 4, and 8).

nk = 2 at layer size [8 1] where the m2 is 96.3%, and the lower model turns out to be na = 1, nb = 2, and nk = 1 at layer size [2 1] where the m3 is 90.75%. **Table 8** shows the comparisons between several different orders and layer sizes, based on the fit between estimation data and simulated output.

The purpose of model validation is to verify that the identified model fulfills the modeling requirements according to subjective and objective criteria of good model approximation. It is important to validate the model using a data set from range (700–1000). **Figure 20** shows the resultant for validation data of three layer sizes at [3 2 2] on model fit.

Time	Orders [nn]			Model fit %		
				Layer size [2 1]	Layer size [4 1]	Layer size [8 1]
	na	nb	nk	zv	zv	zv
Ts = 0.072	1	1	1	85.78	85.88	86.06
	1	2	1	85.71	85.96	85.77
	1	2	2	85.81	86.04	86.23
	2	1	1	87.14	76.86	82.18
	2	1	2	87.13	76.86	82.18
	2	2	1	88.55	72.89	59.69
	2	2	2	88.55	72.54	75.75
	3	1	1	85.95	71	74.4
	3	1	2	85.95	71	74.4
	3	2	2	85.92	80.68	81.27

Table 9.
Comparisons between several different orders and layer sizes, based on the fit between validation data and validation output.

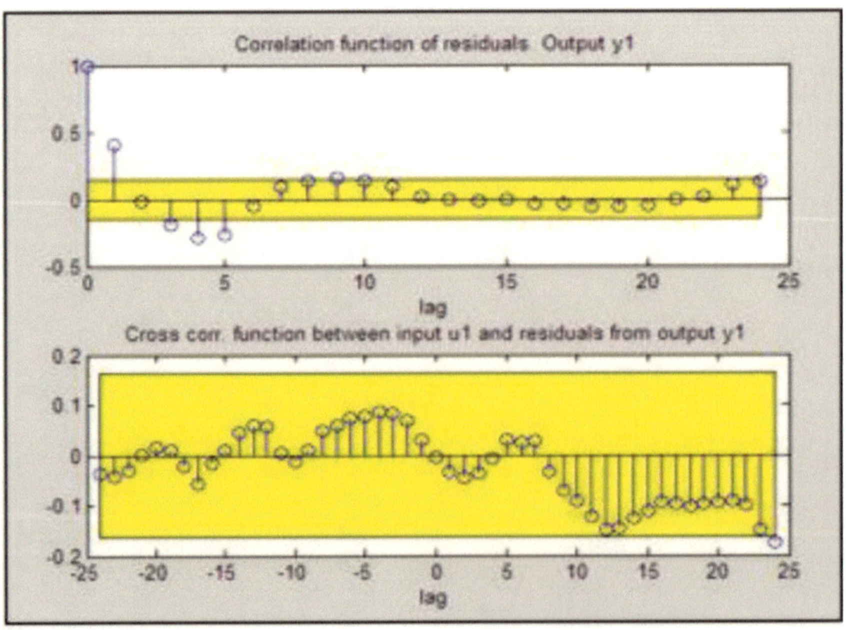

Figure 21.
MATLAB result for residual analysis.

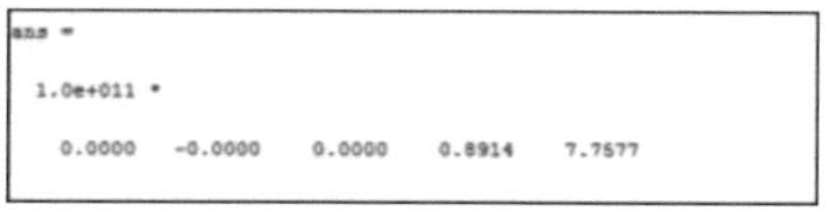

Figure 22.
Weight and bias value at orders [3 2 2].

Based on all the results from the simulation graph, it generally can be said that the model fit is satisfactory in all conditions of training and testing with percentage fit between 60 and 90%. **Table 9** shows the orders consisting of all combinations na, nb, and nk in the range 1–3 and hidden layer effect of the model fit for data validation. The best identified model has na = 2, nb = 2, and nk = 2 at layer size [2 1] where the m1 is 88.55%. The medium model turns out to be na = 3, nb = 2, and nk = 2 at layer size [2 1] where the m1 is 85.95%, and the lower model turns out to be na = 2, nb = 2, and nk = 1 at layer size [8 1] where the m3 is 59.69%. **Table 9** shows the comparisons between several different orders and layer sizes, based on the fit between validation data and simulated output.

This percentage of model fit cannot be used to assume that the model is good. It can be proven when the validation check of this model is done with the residual analysis. The result of residual analysis for this model on validation data is shown in **Figure 21**. For a good model, the cross-correlation between residuals and input does not go significantly outside the confidence region. From the figure, the ARX model has satisfactory correlations. The samples are between the confidence interval.

The weight and bias value at orders [3 2 2] with two neuron in hidden layer [2 1] is shown in **Figure 22**.

6. Conclusion

This chapter has presented the capability of regularized NNARX model structure in representing the system. The nonlinear model of heat exchanger system can be identified by using system identification algorithm. The result shows that the estimated model has a nearly similar trend of output with the output signal of the system based on the percentage of model fit. The major influence to maximize the performance of the model is the selection of proper training data. Overall, with proper training data of regularized NNARX, it is capable in representing the system under test. In conclusion, the result of the best NNARX model of heat exchanger QAD Model BDT921 is na = 3, nb = 2, and nk = 2 with two neurons in hidden layer [2 1]. The model can be written down as:

$$y(t) = 0.8914u(t-1) + 7.7577u(t-3)$$

$$\textit{Loss function} = 0.016355$$

$$\text{MSE} = 0.0162$$

$$\text{ze} = 93.13\%$$

$$\text{zv} = 85.92\%$$

Author details

Tatang Mulyana
Telkom University, Bandung, Indonesia

*Address all correspondence to: tatangmulyana@telkomuniversity.ac.id

References

[1] Tatang M, Mohd Nor MT, Dirman H. A discrete time model of boiler drum and heat exchanger QAD Model BDT 921. In: International Conference on Instrumentation, Control & Automation ICA2009; 20-22 October 2009; Bandung, Indonesia. 2009. pp. 154-159

[2] Tatang M, Mohd Nor MT, Dirman H. A discrete time model of four heat exchanger types HE 158C. In: 7th International Conference on Heat Transfer, Fluid Mechanics and Thermodynamics (HEFAT2010); 19-21 July 2010; Antalya, Turkey. 2010

[3] Dirman H et al. Heat exchanger's shell and tube modelling for intelligent control design. In: International Conference on Computer and Communication Devices (ICCCD 2011), V2-37-41. 2011

[4] Tatang M, Farah Najaa S, Dirman H, Mohamad Nor MT. ARX Model of Four Types Heat Exchanger Identification, MUiCET; 2011

[5] Tatang M, Mohd Nor MT, Dirman H, Afrodi A. Digital Control Design for the Boiler Drum, UTHM FKEE Compilation of Papers; 2009

[6] Tatang M, Mohd Nor MT, Dirman H, Afrodi A. Digital control design for the boiler drum BDT921. In: Proceeding of Conf. on Industrial and Appl. Math.; Indonesia. 2010. pp. 257-260

[7] Tatang M, Mohamad Nor MT, Dirman H, Afrodi A. Digital control design for the boiler drum BDT921. Journal of Mechanics Engineering and Automation. 2011;**1**:392-397

[8] Tatang M, Mohd Nor MT, Dirman H, Hairul A. Modeling and simulation of temperature control PID single loop of a heat exchanger process control training system QAD model BDT921. In: International Conference Electrical Energy and Industrial Electronic Systems (EEIES); 7-8 December 2009; Park Royal Penang, Malaysia. Re: 73-85976. 2009

[9] Tatang M, Hairul A. Modeling and Simulation of Temperature Control PID Single Loop of a Heat Exchanger Process Control Training System QAD Model BDT921, UTHM FKEE Compilation of Papers. 2009. pp. 227-236

[10] Tatang M, Mohd Nor MT, Dirman H, Muhaimin I. Simulation of various PI controllers setting for boiler drums T11 QAD MODEL BDT 921. In: IndoMS International Conference on Mathematics and its Applications 2009 UGM, 034/Mulyana/IICMA/2009; 12-13 October, 2009; Yogyakarta, Indonesia

[11] Tatang M. Parametric and Nonparametric Identification of Shell and Tube Heat Exchanger Mathematical Model [A thesis submitted in Fulfilment of the requirement for the award of the Degree of Philosophy of Doctoral in Electrical Engineering]; Faculty of Electrical and Electronic Engineering Universiti Tun Hussein Onn Malaysia; 2014

[12] Tatang M, Nonparametric A. System identification based on transient analysis with plant process of heat exchanger as study case. International Journal Of Innovation in Mechanical Engineering and Advanced Materials. 2015;**1**(1):19-26

[13] Tatang M, Mohd Nor MT, Dirman H. Parametric Model of Laboratory Heat Exchanger, Keynote Speaker: International Conference on Vocational Education and Electrical Engineering (ICVEE). 2015. pp. 25-37

[14] Tatang M, Judi A, Ekki K. Data analysis using system identification

toolbox of heat exchanger process control training system. In: Fourth International Conference on Information and Communication Technologies (ICoICT). 2016. pp. 466-471

[15] Tatang M. NNARX model structure for the purposes of controller design and optimization of heat exchanger process control training system operation. AIP Conference Proceedings. 2017;**1831**: 020040